David Carpenter

BELOW THE WATERLINE

D1340151

BEARS

HIDE

Published by
Bears Hide Publishing
2 Bramber Avenue
Peacehaven
Sussex
BN10 8LR

First published 2010.
This paperback edition first published 2012.

Also by Dave Carpenter 'Dockland Apprentice' (2003)

ISBN 978-0-9546488-3-1

Printed in Great Britain by
Ashford Colour Press
Unit 600 Fareham Reach
Fareham Road
Gosport, Hants PO13 0FW

Dedicated to all seafarers
Past present & future

Contents

List of plates

Appendix I

Appendix II

Acknowledgements

Richard Joseph — Publishing assistance (www.rakaia.co.uk)
Paul Wood
Brian Anderson — N.Z.S.C. & Shaw Savill
John Layte — N.Z.S.C.
Dave Lang — N.Z.S.C. & British India
Bernard Copleston — Friend & Critic
Coach House Publications — Photograph page 25
Daniel Carpenter — Technical support
Bill Cakebread — Photograph page 242
Southern Counties Radio
New Zealand Shipping Company (N.Z.S.C.)

Cover Picture from a Painting By Robert Lloyd

M.V. *Rakaia*.

Introduction

Some years ago while being interviewed on BBC Southern Counties Radio about my previous book 'Dockland Apprentice', the presenter said "I understand that after completing your apprenticeship you went to sea as an engineer; it must have been like going on a luxury cruise. Whatever did you find to do all day?"

Lots of answers flashed through my mind, but in the short time at my disposal it was impossible to convey what life was like for a young ships engineer in the halcyon days of the Merchant Navy during the middle of the last Century. I simply said, "We worked hard and played hard"!

Instantly the idea for this book came to me.

There have been many books written by seafarers, but very few by ships engineers, so I hope that the following narrative will help to show what life was like for an engineering officer on a motor ship during the early 1960s.

Dave Carpenter, 2010

1
River Mist

What is it that makes a young man want to go to sea? I have heard it said that if you were born in the British Isles then it's in your blood.

I was born in Greenwich, just a stones throw from Deptford Creek, the home of the General Steam Navigation Company, one of the world's oldest successful shipping companies, so perhaps in my case the mist from the River Thames cast its spell on me from the very beginning.

The Second World War had just started, so I grew up with a background noise of whistles and fog horns from ships as they manoeuvred to enter the docks, this was interspersed with the sound of air-raid warning sirens, followed by the hollow crump of anti-aircraft fire from the gun batteries on the high ground of Bostal Heath, Plumstead Common, Woolwich Common and Blackheath. This area was a prime target for the Luftwaffe and the distinct bends in the river at this point gave a positive position for their bombers. On the night of September the 7th 1940, 10,000 were made homeless, 430 were killed and 1,500 were injured in the area around the docks.

Because of the bombing and the need to keep close to our bomb shelter, I was seven years old before I was able to see the great ships in the London Docks at close quarters. Up until then I had only seen the masts and funnels among the towering network of cranes from a distance.

That first visit to the Docks left a lasting impression on me, it seemed that it was the gateway to the world; all I needed was the key to unlock it.

My cousin, Peter Base who was ten years older than me, occasionally gave me books about ships and told me that when he finished his apprenticeship with J&E Hall of Dartford, he was going to join the Merchant Navy. He eventually joined P&O and became a Chief Refrigeration Engineer on many of their great ships.

There was also an organisation for school children called The British Ship Adoption Society organised by The Royal Geographic Society together with the Ministry of Education. Its aim was to encourage association between ships and their crews, with schools throughout the country, over 800 schools and 1,000 ships were members and it gave us youngsters the opportunity to visit our adopted ship and get an idea of shipboard life. The visits were reciprocated by crew members, who gave talks about their ships and voyages. While at sea the ships also corresponded with their adopted school. The scheme inspired many a young man to take up a sea going career.

As the years passed I acquired a liking for all things mechanical, one of my favourite pastimes was watching the steam engines on the Woolwich ferries. An excellent view of these could be had through the large window like apertures around the engine room casing, here the inviting sound of the telegraph ringing and the intoxicating smell of hot steam and oil wafted out into the interior passenger accommodation.

My father had a complete set of 'Shipping Wonders of The World'; I used to spend hours digesting their contents, especially the chapters on the history of the marine engine. All these things, together with the knowledge that my great uncle, Sam Cook, had been a chief engineer in the early days of steam probably influenced my desire to go to sea as an engineer.

While still at school I managed to get a Saturday job with the R.A.C.S. (Woolwich Arsenal Co-operative Society) in Woolwich doing a bakers round with a horse drawn van, I was also able to work during the school holidays. With the money I saved, I was able to buy a motorcycle on a hire purchase agreement. I mention this because in a roundabout way it helped me get to sea.

Apprenticeships, especially in marine engineering, were hard to get during the 1950s, as with many things in life the old saying, 'It's not what you know, it's who you know' rang true. I realised that to get accepted into a ship repair firm in London's dockland, you needed to have relations working

12

there, or to have had your name put down from a very early age, some were put on the list as soon as they were born. I had applied to the well known firms, such as Harland and Wolff and R.H. Green and Silly Weirs who replied that they were sorry, but they had their full intake of apprentices for the foreseeable future. It was at this point that fate took a hand as in my road lived several people who worked in the docks. I became acquainted with one of these because of my motorbike, which was an early racing 'Norton International' that went like a bat out of hell and sounded magnificent. I used to let him go for a spin on it now and again and he always returned with a huge smile on his face, saying "I wish my wife would let me have one of these."

One day I mentioned to him that I was trying to get an engineering apprenticeship in the docks and not having much luck. He smiled and said, "I often have to visit a yard over there; I will see what I can do."

I later discovered that he was an Admiralty Inspector (Surveyor). A couple of weeks later I received a letter from the London Graving Dock Company requesting me to attend their office on the Isle of Dogs (now known as Canary Wharf) the following Monday for an interview. Consequently

The Author on his much modified 1933 Norton International (circa 1956).

13

I was indentured to them for five years, which gave me the necessary qualifications to join the Merchant Navy as a Junior Engineer. In those days to become a ships engineer you had to have served a five year apprenticeship in heavy marine engineering. Various shipping companies had their own schemes where they took you on as a company cadet. In some ways these were better, in others they were not. The cadets had the advantage of a better academic education and got a certain amount of sea time in, where as shipyard apprentices usually did one day and one evening a week at college, but ended up with more hands on experience. It was also sometimes possible to get to sea if you had served an apprenticeship in a locomotive works or perhaps Woolwich Arsenal.

Sadly my five years with The London Graving Dock soon came to an end. I had been taught many skills, some not connected with engineering, but very useful when working in the docks! The work force, were a good crowd of people, with many great characters, some of whom were eccentric to say the least.

Their language also rubbed off on me. No not the swearing! But the terminology used on ships and in engine rooms. I learnt that the ceiling was the deck head, the cabin floors were ceilings, left and right were port and starboard, front and back, were fore and aft and walls were bulkheads. There were no such things as stairs and if there were, you never went up or down them, only topside or down below. Down below in the engine room you never rebuilt a piece of machinery, you always boxed it up! Big and little ends became bottom and top ends and on motor ships the cylinders were called units.

Unlike modern ships engines which are designed for the use of 'hydraulic spanners' we had to know how to swing up to 56lb hammers with confidence, heavier ones were known as monkeys – they resembled the type used today by the police for knocking down doors during a raid. Engine room monkeys were larger and were suspended from a length of rope or chain and swung against a large spanner.

By the time my apprenticeship finished all these things and more were second nature to me, so that when I eventually

14

went to sea I was well versed in ship board language. I have to say that I was never asked to go to the stores for a left handed hammer or a long weight or even make the tea. In the beginning, some errands made me suspicious, such as 'see if you can find a one inch hollow punch' or 'see if you can borrow a fockswedge'. The dockland community that bordered the River Thames was very close knit and family orientated, so I considered myself fortunate to have been accepted. When I completed my apprenticeship in 1960 and commenced my sea going days the London Docks were at their peak, handling 60 million tons of cargo in that year alone. No one could have imagined that the writing was on the wall for the complete transformation of London's Dockland and its workforce in the not too distant future.

2
Spoilt for Choice

During my five years with the London Graving Dock I worked on many different shipping company's vessels, so I had a good idea of what working and living conditions prevailed aboard them. The tramping company of Watts Watts & Co were probably top of my list of going to sea with. Their ships had excellent crew accommodation and were maintained to a high standard.

Bottom of the list were 'Hogarth & Sons Ltd' their ships seemed to be run on a shoestring. Because of their frugal ways they were known as 'Hungry Hogarths'. Their vessels were, in the main powered by triple expansion steam engines and while always interesting to work on, they were not where the future lay. Some of the biggest company's whose ships we regularly worked on were Ellermans; they operated around ninety ships in the late 1950s. At that time their vessels ranged from the old 1929 turbine powered *City of Lyons* to the magnificent sisters, *City of Port Elizabeth*, *Durban* and *Exeter*, all Doxford powered. Then there was Port Line with thirty two ships, and Shaw Savill with thirty, among which was the 26,463 tons gross *Dominium Monarch* powered by four Doxfords and the largest motorship to regularly visit the London Docks. They were all high on my list as they travelled far and wide.

In the 1950s and early 60s, the only way to visit the places with strange sounding names that I had seen on the sterns of those huge ships on my first visit to the docks was with the Merchant Navy. But things didn't quite turn out as I expected!

On finishing my apprenticeship I decided to take a short holiday on the island of Jersey. While there, fate took a hand and I ended up taking a job on a farm for five months. This delayed my getting to sea, but I did learn how to grow tomatoes and plough a straight furrow. Unfortunately the farm has long since disappeared under the sprawling development of

St. Helier. Due to the night life and abundance of beautiful girls I found I was suffering from financial cramp, so it became imperative to get to sea as soon as possible.

I returned home to Woolwich and decided the quickest way would be to go up to the shipping company's offices in the City. Most of them were situated in Leadenhall Street and had very imposing and somewhat daunting façades, but at the same time, drawing you to their windows to drool over the magnificent models of some of their company's ships.

Strolling down Leadenhall Street trying to make up my mind where to try my luck, I suddenly found myself outside the large entrance to number 138, the offices of The New Zealand Shipping Company.

I have no idea why I entered their premises, I had never worked on any of their ships, The Graving Dock had nothing to do with them and all repair work on their ships while in London's Royal Albert Dock was carried out by the ship repair firm of R.H. Green & Silley Wier. I knew their ships by sight, with their pale yellow funnels, white topsides and black hulls, they had a yacht like quality and gave me the impression they could take anything the sea could throw at them.

As I entered the reception area I was immediately confronted by the concierge or doorman. He had his hands behind his back and had a long square sided face and large hawk like nose which seemed to be attached to the peak of his hat.

"Can I help you young man?" he asked as he looked down each side of his nose at me.

"I've just popped in to see if the company has any vacancies for a junior engineer," I replied.

"Have you an appointment?" he asked.

I said I hadn't and in a typical Prima Donna attitude he told me that they had none.

'Fair enough', I thought, I turned and retraced my steps to the pavement.

Across the road I could see the offices of The Blue Star Line, so I thought I would give them a go.

While I was waiting to cross the road, I heard a shout and

looking round noticed the Prima Donna running towards me.

"Did you say you were looking for a junior engineer's position." He gasped.

I told him I was going over to the Blue Star office.

"No you must come with me," he almost shouted, as his hat tipped back on his head, proving that it wasn't attached to his nose after all.

I followed him back to the building and into the reception area, where he told me to wait. After a minute or two a young office worker appeared and asked me to follow him. On the way up an elaborate flight of stairs he asked me my name before coming to a stop at a door marked Personnel, he knocked, opened the door and said.

"Mr. Carpenter is here sir."

From that moment and the rest of the day my feet hardly touched the ground. After a short interview with the personnel officer, I was asked about my qualifications. I produced my indentures, which he studied for a minute or two and said.

"That's good enough for me, I want you to take this note and go round to the P&O buildings just down the road and have a medical. If you pass, come back here and I'll give you a note for the Shipping Federation, they will give you a Discharge book, a British Seaman's ID and a classification certificate."

I eventually found the medical officer somewhere in the inner sanctum of the P&O buildings and gave him the note. I had never had a medical in my life, so I was rather concerned about what he might find wrong with me. I need not have worried, I passed with flying colours; the note, no doubt having something to do with it. I was also given a polio jab and a vaccination certificate and told to keep it in my discharge book that would be issued to me at the Shipping Federation.

I returned to the personnel dept; where I was given another note, (already prepared) together with directions to the Shipping Federation. Once there, I was asked various questions, then told to go next door to have my photograph taken, if my memory serves me right, it was in the back room of a newsagents.

I was then told to proceed to Oliver's, the marine outfitters, to be kitted out with the required uniform. This was quite extensive, consisting of a blue uniform, dark blue Burberry style raincoat, black shoes, socks and tie, plus the tropical whites consisting of two white short sleeve shirts, two pairs of white shorts and long white socks and a pair of white shoes and two white boiler suits.

In those days, like all young men, I was a dedicated follower of fashion and always had my suits made to measure, so I chose to have my blue uniform made the same way. On the recommendation of the tailor, I had it made of doeskin. This proved to be good advice, as doeskin, under normal conditions never seems to wear out, I still have it and it still fits. (Just!)

The whole outfit came to about £50, a lot of money in 1960, especially as a junior engineers salary was £47 a month. Fortunately 'Oliver's' had said they had an arrangement with the N.Z.S.C. It meant I didn't have to pay there and then, as it could be deducted out of my future wages. Never having worked on New Zealand Shipping Company's ships during my apprenticeship, I was surprised to learn from the tailor that within the N.Z.S.C. the gold insignia of rank was not worn on the sleeves, but on the shoulders in the form of epaulets, so I had to purchase two pairs, one for the blue uniform and one for the whites.

Engineers all had gold on purple background, the juniors had one narrow gold stripe, the 4th one wide stripe, the 3rd, two, the 2nd three and the chief had four. While in the City I also had to obtain a passport, which was no trouble as I now had my British Seaman's ID book.

After all this I returned to Leadenhall Street, where I was given another note and told to go to the Royal Albert Dock and report to Mr. Strachan the marine superintendent. I was shown into his office, where he told me I would be joining the M.V. *Rakaia* as its junior engineer at the beginning of January, in the Port of Avonmouth. The good news was that

I was put on the payroll straight away, with an advance on my wages and a first class train ticket to Avonmouth.

I returned home hardly believing my good luck, after all the years since my first visit to the docks, when I saw the gateway to the world, I had finally got the key. I immediately looked through my latest copy of Lloyd's List, I always bought one every Wednesday! Quickly studying the N.Z.S.C. ships movements. I discovered that the *Rakaia* was discharging part of her cargo in Antwerp and would be arriving back in the U.K. in a few days, to complete discharging and then load a general cargo in Avonmouth for Australia.

3
Joining Ship

The day of joining ship soon came round; I had collected my sea going kit from Oliver's the previous week, just before they closed for Christmas. It all looked very impressive, but I did think I looked a bit odd in the tropical whites. How wrong can you be, when the time came to wear 'Whites' it went quite naturally and I soon found out that there was nothing as smart as a good looking ship with her officers dressed in whites.

I decided to wear my uniform for the journey to Avonmouth; it shows how green I was! I later discovered that Merchant Navy officers very rarely wore their uniforms ashore. It was different for the company Cadets; they were expected to wear theirs ashore.

I arrived at Paddington Station with my gear and as I approached the platform to board the train, a group of lads in uniform came over and asked me if I was joining the *Rakaia*. Being January, I was wearing my blue raincoat; therefore they were unable to see my single gold stripe, so they took me for a cadet and were endeavouring to make me welcome.

When I explained that I was joining the *Rakaia* as a junior engineer, they lost interest and walked away. I was beginning to get my first indications of shipboard etiquette and pecking order.

After a long and uneventful journey, I arrived at Avonmouth in the late afternoon and took a taxi for the short ride to the docks.

I soon recognised the unmistakable shape of a N.Z.S.C. boat ahead of me, by this time it was getting dark, but as I approached I could make out the name RAKAIA and underneath LONDON on her stern.

I had spent the last three years of my apprenticeship working on all builds of ships throughout the whole London dock system. By the necessity of saving myself very long walks, going down the wrong side of a dock looking for the next ship I had to work on, I had learnt to recognise various ships from a distance.

From 'The Cut of Her Jib', with the shape of her rather squat funnel and cruiser stern I thought she was probably built by Harland & Wolff.

At last I was about to board my first ship as a sea going engineer! As I approached the steep gangway, I had to dodge out of the way of the electric trolleys stacked with boxes that were keeping the Dockers fed with a constant assortment of cargo from within the adjacent warehouse.

There was lots of activity; both on the quayside and the ship, the cranes were swinging back and forth, some delivering large wooden crates while others lifted great bundles of boxed goods in cargo nets.

This was an environment that I had become used to, so I was feeling quite at home as I climbed the gangway. I left my gear just inside the accommodation while I went in search of the Chief Engineer's cabin to report my arrival. After the cold and damp quayside, the accommodation felt warm and friendly, with an inviting smell wafting out of the adjacent pantry. As I entered the engineer's alleyway, it blended with a faint aroma of hot oil and fresh paint. The Chief's cabin was on my immediate right, the door was open, so I knocked and looked in, he was laying on his daybed reading a paperback novel. He waved me in and I explained who I was.

After welcoming me aboard he explained that he was the stand by Chief and that the sailing Chief was expected back from leave in the morning, adding;

"The junior's cabin is down the alleyway opposite the engine room door, dump your gear, then have a look around down below, you'll find some of the lads finishing off for the day down there, by the way, dinner's at five in the saloon."

I collected my bags and went down the alleyway passing the 2nd, 3rd and 4th cabins and so on, until I came to the one marked 7th Engineer, the next one was the Junior engineers, a quick look further on told me the rest of the cabins were the 2nd Electrician, Chief Electrician, 2nd Refrigeration Engineer and lastly at the after end of the alleyway was the Chief Refrigeration Engineer.

On entering I took in at a glance its layout, there was a port light at the far end; beneath it was a bunk which took up almost the full width of the cabin. To my left was a small wardrobe and between it and the bunk a wash hand basin. On the right, against the forward bulkhead, was a desk with a fold down front, then a daybed with a fitted cover in some sort of green floral design. There was also a comfortable looking chair with matching cover. In all it was about ten by eight foot, quite adequate for normal purposes. I later discovered that all the engineers' cabins were the same, except for the Chief, 2nd and 3rd and the Chief Freezer's which were slightly larger. I took off my raincoat and jacket and put on one of my gleaming white boilersuits, stepped across the alley, opened the heavy steel door and entered the engine room.

I found myself on a small grating at the top of a flight of steps about twelve feet above the after end of the main engine. I had no idea of what type or make of engine to expect, but I immediately recognised it as a Burmiester & Wain double acting two stroke made by Harland & Wolff (Belfast). I was somewhat pleased about that, because I had worked on quite a few of this type during my apprenticeship, so I was quite familiar with their internals. Operating and maintaining them under seagoing conditions was another matter.

This was the view on entering Rakaia's engine room. It shows the top exhaust piston yokes with the exposed exhaust pistons beneath and the coupling rods to the bottom exhaust pistons. This picture by A. Cunningham is reproduced from 'Empire Food Ship's' by Richard P. de Kerbrech and shows one of the engines of Shaw Savill's M.V. *Waiwera* but is exactly the same as the *Rakaia's*, both ships were built at Harland & Wolff's Belfast Yard and launched in 1944.

The *Rakaia's* was an eight cylinder (unit) engine with the exhaust pistons driven by large eccentrics from the crankshaft and was an improved design on the earlier pre-war version, where the exhaust pistons were driven via links from an auxiliary crankshaft.

As the main purpose of this book, is to try to convey what shipboard life was like for an engineering officer during the middle part of the last century, I have left the more technical details until the appendices at the end.

I descended to the next level, which gave access to the top exhaust pistons, indicator cocks, fuel and air start valves. The exhaust pistons were carried by huge yokes which rose and fell in tandem with the lower ones with every revolution of the crankshaft. When under way, all eight of them went up and down in a seemingly disordered formation among a haze of hot oil and diesel fumes. It was not the place for the faint hearted!

Sometime ago I spoke to an ex deck cadet who sailed on the *Rakaia*. He remembered having to enter the engine room with the engine running at full sea speed; he described it as the most terrifying place he had ever been in.

At the far end, beneath the funnel, on a wide platform was the boiler, force of habit made me glance at the large pressure gauge mounted high above the furnace door. I saw that the needle pointed to 90lb /sq in and the water gauge showed three quarters full.

The boiler was mainly used for accommodation heating and keeping the engine cooling water warm when in port. When at sea, steam was generated by some of the waste heat from the main engine exhaust; this was diverted according to demand via a large flap valve in the exhaust trunking. It was one of the watch keeping junior engineer's duties to maintain the pressure and water level when under way. When in port fuel oil was used to raise steam and it became the responsibility of the Donkey Man, under the watchful eye of the duty engineer.

I could hear an engine running and as I descended to the next level, which gave access to the camshaft and fuel pumps, I realised that it was coming from one of the ships four generator's way down at the bottom of the engine room. The next level down gave access to the bottom exhaust pistons, main piston rod stuffing boxes, lower fuel valves, scavenge drains and indicator cocks.

As I went down the last flight of steps, I could hear the familiar 'chinking sound' of spanners being put down on engine room plates, so on reaching the control position, I made my way in the direction of the sound which led me to the forward end of the main engine. Here I found two engineers making adjustments to one of the two boiler feed pumps.

These together with a ballast and emergency bilge pump were the only auxiliaries that were steam powered. There was an evaporator but it was well past its sell by date and hadn't been used for many years. As I approached, they looked up and greeted me with hand shakes. They introduced themselves as Jim the 'Fiver' and Norman, the 'Sixer'.

They spoke in very broad Scottish accents, which I had difficulty in understanding. I was familiar with a toned down version of the Scottish tongue, as during my apprenticeship, I had worked with many fitters and boilermakers who originated from that part of the world. These had all lost some of their broad accent after many years in the South.

Today we are all familiar with regional accents, mainly due to the television, which now depicts the differing accents in its programmes.

During the 1950s, everyone on the radio and television spoke with a BBC voice, even someone playing the part of an East End whore had a voice comparable to that of a news reader.

The two lads in front of me were almost speaking a foreign language!

They in turn were having trouble understanding my Cockney accent. Despite these minor problems, we got on well and as time went on we became firm friends.

They gave me a quick tour of the engine room, showing me where most auxiliaries were, including the air bottle flat, where the two huge air receivers were. These were used to start and manoeuvre the main engine and for starting the generators. Next to them was the electrician's store and fridge machinery room which was a temple to cleanliness. It was absolutely forbidden to cross its threshold wearing engine room shoes.

This was where the gleaming compressors made by Messrs J&E Hall were tended like queen bees by their two and sometimes three refrigeration engineers in their gleaming white boiler suits.

Across on the Starboard side was the engineer's stores and workshop; this contained a very old lathe and pillar drill. There was also a long steel bench with a large vice and an electric grindstone. Just outside the stores, taking up practically the remaining width of the engine room was the main electrical switchboard, with its four huge contact breakers, ammeters and voltmeters. It all looked very daunting, but I new I would have to master its intricacies before I was much older.

The Author as a very green Junior Engineer (1961)

4
Part of the Crew

After my quick tour, the fiver said it was time we went up and got ready for dinner and invited me to join them that evening for a few beers in a pub that they knew somewhere up river. I had a shower in the engineers wash room which was opposite the 2nd electricians (lecky) cabin and presented myself in the officers dining saloon. This was at the forward end of the engineers alleyway, but on the starboard side of the cross alleyway. On the port side was the officers smoke room. One of the stewards showed me where to sit, as the tables were arranged in a certain pecking order.

At the centre table sat the Captain, Chief and 2nd Engineers, 1st Officer (Mate), 2nd Mate, Chief Radio Officer (Sparks), Chief Steward, Surgeon or 'Doc' as he was called on board and Schoolmaster (Schooley). At the forward table, Portside, sat the 3rd and 4th Mates and the 2nd Sparks. On the Starboard forward table, the 3rd and 4th Engineers, Chief Refrigeration (Freezer) Engineer and the Chief Electrician (Lecky). The 5th, 6th, 7th and Junior Engineer, together with the 2nd Lecky and 2nd Freezer, sat at a large table on the after Portside.

There was a small table starboard aft, reserved for officers and their guests.

The reason the *Rakaia* carried a Doctor and a Schooley (who had a 2nd Mates ticket), was because she was a Cadet Ship, with around thirty Cadets. When at sea, uniforms had to be worn at all times in the saloon, while in port, any officer going ashore after meal times, could wear a suit and tie with the Captain's permission. Etiquette and protocol were strictly adhered to, smoking was not allowed and noisy conversation and messing about were highly frowned upon. The only person who could raise their voice in the saloon was the Captain, this he did on occasion when he had a particular point to make that he wanted his officers to hear, or about somebody's dress code.

The menu varied from day to day, certain popular items would be repeated on a monthly basis, the food during my time on the *Rakaia* was always first class. This was entirely due to Jerry the Chef and his galley team of two cooks, Butcher, Baker and Galley Boy, overseen of course by the Chief Steward.

My first meal on board the *Rakaia* was a completely new experience for me. Previously most of my meals had been at home or while at work, taken in pubs or smoke filled dockside cafes. This was in the company of unscrupulous looking but good hearted characters whose sole intention was to eat their meal as quickly as possible, light up a cigarette or pipe, clear away the table and start the obligatory card school. I soon came to realise that I was entering a different way of life, meal times on board ship were the main social times of the day. This was especially so in port, when all officers would normally be present.

After the Steward showed me to my seat, Jim the 5th Engineer introduced me to the rest of the lads at the table. After dinner we excused ourselves and went into the smoke room where we were joined by several other officers, including the 3rd and 4th Engineers. While working on ships as an apprentice, I looked on Engineers of such rank with respect, these together with the 2nd Engineer were senior officers and during their watch at sea, had the full responsibility of the engine room in their hands.

There were six watches in a twenty four hour day:
The 2nd Eng. took the 4.00 am – 8.00 am then the 4.00 pm – 8.00 pm, assisted by the 5th;
The 4th Eng. the 8.00 am – 12 noon then the 8.00 pm – 12 midnight, assisted by the 7th;
The 3rd Eng. 12 noon – 4.00 pm then the 12 midnight – 4.00 am, assisted by the 6th.

Under normal circumstances the electricians were on day work, as were the freezers and junior engineer when outward bound. Once frozen cargo came aboard, the junior joined the freezers and they went on to four hours on eight off watches.

The Chief Engineer didn't take a watch, but was on call at all times should he be required.

After introductions and a chat with some of the officers in the smoke room, Jim suggested that we go ashore as planned for a few beers. The reason behind this was that the bonded stores would not be available until after we sailed, so the ship could be considered dry until we got to sea.

I can't recall much about my first night ashore. I remember we had a taxi to somewhere up river, where we then clambered aboard a rowing boat that was some sort of ferry. The boatman then took us across the river to a muddy slipway next to a pub. The rest of the evening soon became a blur!

The next morning the engineer's steward woke me at 7.00 am and said that breakfast would be at eight. Over breakfast a post-mortem was held on the previous night's escapades and it was concluded that a good time was had by all. Although I had a stinking hangover I prided myself that I was now one of the *Rakaia's* crew.

When in port, we went on day work and each engineer, except the Chief and 2nd took turns being duty engineer for 24 hours. This mainly meant making sure that the engine room behaved itself while everyone was ashore or asleep and being available to advise any shore gang or visitors. In reality it required regular visits to check that the generators weren't developing any problems and to keep an eye on the pressure and water level in the boiler. This was looked after by the Donkey man, who also acted as Storekeeper, but an independent check by the duty engineer tended to give a belt and braces effect against anything untoward happening.

When on day work we started at 9.00 am and finished at 4.30 pm with an hours break for lunch. About 10.30 am and 3.00 pm we would have a tea break, this was known as 'Smoko' and would be taken in a small mess room at the top of the engine room on the Starboard side of the main access doors. These were always kept open, except in heavy weather, when it was possible for rogue waves crashing on the after deck to empty themselves into the engine room.

On any ship, the last 24 hours before sailing was always a hectic time, sometimes bordering on panic, with many unfinished, 'Just Needs', jobs requiring completion. Sometimes pistons or bottom ends waiting to be boxed up. Such was the situation when I turned to that first morning. The shore gang were still busy working on one of the four generators and in the main engine crankcase after fitting new rings to three of the main pistons.

Because of shore side union rules, ships engineers were not allowed to do repair work while in U.K. ports. We were only permitted to do routine maintenance and emergency repairs. It was a different matter abroad, we were expected to do all the repair work ourselves, sometimes however, when a lot of work had to be done and time was of the essence, a shore gang came aboard to assist us.

I know for a fact that many ships ignored the rules in the U.K. and re-ringed pistons etc. during the night. If the unions got to know about it, they would call all their members out on strike until the matter was resolved.

I soon realised that the three most useful pieces of equipment that a Ships Engineer needed to carry were a lump of rag, a wheel key for opening tight valves and a torch, very useful for identifying individual pipe runs in the tangled complexity of the mass of pipe work to be found in all engine rooms.

I knew I would have to familiarise myself with all the different systems, such as fuel, air, lubricating oil, bilge and ballasting as soon as possible. In an emergency I might need to find any one of them in the dark. Wheel keys and fire extinguishers were always kept at strategic positions around the engine room for just such occasions.

When we knocked off for lunch, I had a visit from the sailing Chief; this was John Cowper, an elderly Scotsman in his early to mid sixties. As time went on, he was a person I looked up to with high regard. He was an engineer of enormous experience; he treated his engineers as if they were his own sons and would have nothing said against them by the deck mob.

The purpose of his visit was to introduce himself and to tell me to move all my gear into the 7th engineer's cabin.

"I see you served your time with the London Graving Dock, with your experience I would sooner have you in the engine room than in the freezer department, you'll be on watch with the fourth on the eight to twelve," he remarked over his shoulder as he went back up the alleyway.

Many years later I discovered that when he went home on leave, he visited the shipyards in his area and recruited lads who were near the end of their apprenticeships into the New Zealand Shipping Company. This would explain the predominance of Scottish engineers aboard the *Rakaia*.

I had only been on board for a day and already I had been promoted to 7th. Engineer!

Unfortunately it didn't come with a rise in pay! All engineers up to the Fourth received the same, i.e. £47 a month plus Sundays at sea and a 'section A bonus'. It was called this because when we signed the ships 'Articles' it was under the 'Section A' agreement. It basically meant that we didn't receive payment for overtime, so in theory we were under obligation to work every hour for the next two years or until the ship returned to the U.K.

Chief Engineer John (Jock) Cowper.
(Photo Paul Wood).

In reality most voyages (trips) lasted about four months and if everything was running well when under way, we worked four on eight off and in port usually from 9.00 am to 4.30 pm. If we had problems, then we worked around the clock to get them sorted, this was what the 'Section 'A' bonus' was supposed to compensate for.

One of the good things about the bonus and also the Sundays at sea was that we could not sub on them during the voyage. This meant that if we spent all our wages on cigarettes and beer from the ships bond and in foreign taverns of ill repute, we would still have some wages to pay off with at the end of the voyage.

The next morning all the sailing crew having joined the ship, we had to go into the saloon and sign the ships articles. We were supposed to read the articles of agreement before signing; this was a throw back to the old days of sail, where unscrupulous ship owners could dupe the crews into signing on before the mast with appalling conditions of food and accommodation.

We never did read it, if we had it would have listed the minimum amount of different foodstuffs and minimum size of accommodation that we were entitled to under the Merchant Shipping Act. Within the N.Z.S.C. the food was always good and on the *Rakaia* sometimes surpassing what might be termed as excellent and the accommodation was clean and adequate.

It was also inspected every week by the Captain and his entourage to make sure certain standards were maintained. Although the conditions down in the engine room were accepted as quite normal, in today's climate of health and safety they would not be tolerated. The heat, noise and exposure to oil, carbon and asbestos among other things would make it a no go area.

Later in the day we had a fire and boat drill, this made sure we all knew where our stations were and what we had to do in the event of an emergency. Life jackets were kept under our pillows and they had to be worn during boat drill, commonly known as Board of Trade Sports. They were the Kapok filled

34

type and very cumbersome, anyone jumping overboard wearing one had a good chance of breaking their neck when they hit the water if they didn't hold them correctly.

After boat drill, we all turned to down below getting everything ready for sailing. We were due to sail that evening at 8 o'clock, a notice was put at the top of the gangway, saying, 'All Shore Leave Ends 1600hrs'.

5
Preparing for Sea

It's not possible to explain, but I seemed to feel a different sort of atmosphere envelope the ship, every one spoke in lower tones, but with a sense of purpose, all the hammering and rattling of chain blocks had disappeared along with the shore gang, although one of the generators was running, the engine room seemed strangely quiet and empty. All the rubbish left on deck by the Dockers had been cleared and all the hatches and derricks had been secured by the Cadets under the watchful eye of the Bosun and Second Mate.

We had our dinner as usual at 5.00 pm then returned to the engine room.

During the course of the day, I had learnt that our destination was Sydney, Australia. We were going through the Mediterranean, through the Suez Canal and down the Red Sea to Aden to take on fuel, over the Indian Ocean to the West coast of Australia, then across the Great Australian Bight, round to Sydney.

After loading at various ports on the coast, our route would be across the Pacific, making a short stop at Pitcairn Island, before going through the Panama Canal to the Dutch West Indian Island of Curacao for bunkers. After discharging part of our cargo to various ports on the Eastern freeboard of North America, we were to go on to Canada and up the St. Laurence River to Montreal. We would then head across the Atlantic to Hamburg and Antwerp, before discharging the remainder of our cargo at London's Royal Albert Dock. This meant that on my first trip to sea I would be sailing right round the world, plus a few diversions along the way, a total distance of about 30,000 miles.

The 2nd Engineer put me in charge of the two air compressors, stating;

"Keep your wits about you it's going to be a long night!"

Once the ship started manoeuvring out of the dock they would be going flat out, trying to keep the air bottles supplied depending on the amount of starts of the main engine. I also had to keep a close eye on the temperatures of the engine oil and water jacket coolers and adjust them accordingly. Keeping the water jacket temperature steady was of particular importance; sudden changes could cause the main engine liners to crack, with disastrous results.

I was told that as we had several engineers who were not familiar with a Burmiester and Wain main engine, everyone except the Chief and the Freezers would be on duty as we got under way; this was to make sure that the work carried out by the shore gang was working OK.

Perhaps I should mention that the *Rakaia* was the only ship in the company that had a B&W main engine. All the other company's motor ships were powered by Doxfords or Sulzers.

Once everything was settled down and going well, we would go on six hour watches, this meant there would be three engineers on watch with one keeping close to the controls just in case any emergency orders were rung down from the bridge. When well clear of the Bristol Channel, we would start our four on eight off sea going routine.

Finally everything was ready, the steering gear, engine room telegraph and the whistle had been tested and a stowaway search completed, the Fiver had started and put another generator on load, the fuel system had been primed and the air bottles were full and both compressors were running off load. The turning gear was out, a notice near the controls confirmed this had been done and the main valves on the air bottles were open. The Chief Lecky was standing by, ready to note down all the times and orders from the telegraph into the movement book and the 2nd engineer was standing back overseeing the whole operation.

The telegraph suddenly rang and the pointer went round to Stand By.

"This is it, there's no turning back now." I thought to myself.

We had started a voyage that would take us to the far side of the world!

The Third, who was at the controls slowly opened the air demand valve above his head and the system filled with a hollow hissing sound sending the pressure gauge needle round to 350lb per square inch.

The 5th and 6th were in position, one at the forward end of the top gantry and one at the forward end of the bottom gantry. They were waiting for the first movement of the engine. Once the telegraph rang down and the third gave the engine a kick over on air, they would run along the cylinders shutting off the indicator cocks. These were left open to expel any water that might have found its way into the combustion chambers. As water is uncompressible, its presence could cause considerable damage.

After a tense wait of a long three minutes, the telegraph rang and an adjacent red light came on as the pointer went to Dead Slow Astern.

The 3rd moved the large reversing lever across its quadrant and the red light went out, he then moved the fuel lever to 'Start', there was a loud roaring noise as the air was distributed to each cylinder in turn, judging it so that the engine only did one revolution, he then brought the lever back to stop and stepped back to get an unobstructed view

View of Rakaia's main engine controls. (Photo Paul Wood)

to the top of the engine, where about twenty five feet above, the Fiver gave the thumbs up that all was well and just above our heads on the next grating up, the Sixer did the same.

Having received the ok, the 3rd moved the fuel lever forward to 'Start', there was another loud roaring and hissing noise and as the great engine turned over once or twice, he pushed the lever farther on to 'Fuel'.

The huge beast started immediately, I was surprised how quickly it responded, the rev counter shot round to 60rpm before being bought back to 35rpm, which was Dead Slow. Almost straight away 'Stop' was rung down, anticipating the next movement the 3rd moved the reversing lever to 'Ahead'.

The 2nd signalled to me to put the compressors under load, this was to try and keep the air bottles topped up. While I was round the other side of the engine I heard it start up and by the time I returned to check the coolers it was running 'Slow Ahead'. We had several more movements before the bridge phoned down to say that we would soon be dropping off the Pilot. By now the cooler temperatures had risen, it was difficult to keep them steady while manoeuvring was taking place, the gauges were slow to react to the cooling effect of the sea water passing through the coolers, so the temperature reading of the water jackets was off the mark. The 2nd came over and told me in no uncertain terms to get my act together, then went back to the controls where I heard him say to the 3rd something about bloody first trippers.

The valves had to be adjusted in advance of what the gauges read. This took me a while to get the hang of. After a short stop to let the pilot off, Half Ahead and shortly after, Full Ahead was rung down; it looked like we were finally under way. I was now able to keep a better control over the temperatures as they slowly worked their way up. The 3rd gradually bought the revs up to 102 rpm, I later learnt that this was the most economical speed, if increased by half a rev it would increase the fuel consumption by around five tons a day.

The *Rakaia's* engine was unusual in that it used gas oil rather than heavy or crude oil; this did away with the need for fuel heaters and their associated problems. It also had the advantage of cleaner running and less gumming up of piston rings. In average sea conditions, with the engine running at 102 rpm she would use 28tons in 24hrs. This gave her a speed of 14.5 knots.

I could feel the ship beginning to roll slightly, so I hoped I wouldn't be the victim of seasickness! The air bottles were now full and with the ship well under way the 2nd told me to shut down the compressors, (in theory they wouldn't be needed until we reached Port Said), then change the sea suction from high to low, this was to eliminate the possibility of sucking in air to the cooling systems, but to leave the main air valves on the bottles open for a while, just in case of any sudden orders from the bridge.

While all this was going on, the 4th, 5th and 6th and with the junior under their care were busy looking after the generators, fuel purifiers, boiler feed pumps and keeping the bilges dry. An eye also had to be kept on the stern gland, this entailed a long walk alongside the revolving propeller shaft down the shaft tunnel, checking the shaft bearings on route.

The engine room was beginning to heat up and an oily mist was hanging around the bottom exhaust pistons. The 4th came and spoke to the 2nd and they both disappeared round to the other side of the main engine. A few seconds later the fourth retuned and started up one of the two spare generators, apparently an air start valve on number four generator was faulty and its feed pipe was glowing red hot. This potentially dangerous situation sometimes occurred after a generator had been running for an hour or so and was usually caused by a piece of carbon getting stuck on the valve seat during starting. The lads descended on the generator and soon had it ready to start again if required.

We had been under way for three hours when the 2nd decided that as things were going O.K. and we were well clear of land, instead of staying on six hour watches we would start the normal sea routine of four on eight off. The last three hours seemed to have flown by and as I was to be on watch with the 4th on the eight to twelve, I had only another hour to do before I could turn in and get some sleep before going below again at 8.00 am.

Everyone except the 4th, myself and a greaser knocked off, the 3rd and the sixer would have to take over at midnight, so they wouldn't get any sleep before returning to the engine room for their four hour stint. The Donkey man had shut down the furnace on the boiler and the 4th had put it onto the waste heat from the main engine. It would take a few hours to stabilise the pressure by frequent adjustments of the flap valve, which was right up behind the boiler at the base of the funnel. This was one of the things I had to keep a close eye on. The climb from the engine room plates to the flap valve was about 45 feet, by the time you had reached the engine tops you were sweating, but by the time you climbed up to the flap valve the sweat was literally pouring off your body. Up there in the confines of the funnel it was pitch black and the temperature was enough to fry an egg; the combined heat from the main engine and generator exhausts and boiler, plus the soot and fumes made it a very unpleasant place to be.

The lever that adjusted the valve via a ratchet mechanism was so hot that a special pair of asbestos gloves had to be worn; luckily it only required one or two clicks to make the necessary adjustments. By the time you returned to the engine tops your hair, face and neck were streaked in mixture of soot and sweat. Today this exercise would make the perfect slimming plan for anyone wishing to loose weight, this could be one of the reasons why I never met a fat marine engineer.

Before handing over the watch we chalked a notice on the board indicating that the pressure was on its way up or down, eventually it would settle down and need very little attention. By the end of the watch my brand new boilersuit showed very little evidence of once being white.

6

Watch Keeping

The next morning the engineers Steward called me at 7.00 am. I had a shower and as I put on my uniform I had a quick look out of the port light, it was dull and overcast and there was moderate sea running. It looked very miserable, but I consoled myself with the thought that it wouldn't be long before we would be sun bathing out on deck. How wrong can you be? It just proved how green I was. I had to wait until we reached the Indian Ocean and encountered the latter part of the Northeast Monsoon which would give us a good start down to Western Australia.

After a quick breakfast, I returned to my cabin and changed out of my uniform into my boilersuit. We kept our work shoes in a rack just inside the engine room; I had purchased these from the ships slop chest. This was the terminology for the ships stores and it was under the care of the Chief Steward, it was a throw back to the old days of sail, when the Captain purchased a selection of useful items that he knew the crew would need during the voyage. When a crew member bought something the cost would be deducted from his wages when he paid off at the end of the voyage. The difference the Captain paid and sold the articles for, would result in a nice profit for himself.

Our steward had given me two small booklets of chitty's; one was for the Slop Chest and the other was headed Wine List. If I wanted a case of beer I just filled in one of the chitty's and signed it, then left it on the desk in my cabin, the Steward would then deliver it during the morning and just like in the old days of sail, the cost would be deducted from my wages at the end of the voyage.

FORM No S43

WINE CARD

SIGNED........................

DATE...................... BERTH No. ....

ALL ACCOUNTS TO BE SETTLED WEEKLY

I still had a few days to go to cover the advance paid to me when I joined the Company; this time was known as flogging a dead horse, another throw back to the old days. Apparently in those days, after they had worked out their advance they made a horse out of straw and for some reason thrashed it with sticks.

It was nearly five minutes to eight, we had been under way for twelve hours and as I pulled open the heavy steel door to the engine room, the noise and heat hit me like an express train. I stepped in and exchanged my flip flops for my engine room shoes, these were slip-ons with steel toe caps. We never wore socks or lace up shoes; in an emergency, the time lost putting them on might cost us dearly.

As I descended the first flight of steps, I watched the top piston yokes flying up and down through the oily haze that now filled the top of the engine room. The further down I went the clearer and cooler it became. When I reached the controls the 4th was behind me, it was frowned upon to be late, after four hours down below the previous watch would always be looking forward to a few beers. After informing us about anything that needed keeping an eye on, they left us to get on with it.

Our greaser nearly always got down before us; he brought with him a large brown enamel teapot full of steaming hot tea. It had a long oily rope handle making it easier to carry down all the flights of steps. If everything was running O.K. we both had a mug of boiling hot tea, strangely this helped to keep us cool. About half way through the watch he nipped up and made a fresh pot. When the pot cooled down, we boiled it back up again by injecting steam from a pipe connected to the drain on the boiler feed pump. This tended to add a certain flavour, no doubt caused by the boiler treatment chemicals.

During the daytime we had three engine room labourers working with us, their main task was to keep things clean. The design of the *Rakaia's* engine made this virtually impossible, it continuously emitted fine droplets of oil as the piston skirts exposed themselves and every now and again one or more main piston rod stuffing boxes would blow a shower of oil over the immediate area, until sealing itself after a few revolutions.

The loud whooshing noise made by the offending stuffing box could be heard from our accommodation and out on deck. If you happened to be in the vicinity you were instantly covered in black oil.

One of the regular duties the junior engineers had to do was to make sure that the scavenge drains were kept clear. Each unit had a two to three inch diameter drain pipe coming down from the scavenge trunking, on the end was a large valve that was adjusted so that it let the scavenge breathe into an open ended five gallon oil drum.

These pipes regularly became blocked with a carboniferous oily tar like substance, if we couldn't feel the end of the pipe sniffing it meant it was blocked. We had an assortment of rods with various ends that we used to unblock them. This was a particularly unpleasant job, they were situated along a grating that went the length of the engine and had the top exhaust trunking overhead and the bottom one below and the scavenge blowers to the side. It was known as 'Hells Alleyway' because it was very gloomy and the temperature was in the region of 160 degrees Fahrenheit.

If a drain was blocked, it often took sometime to clear it, when it finally cleared, it would do so with a huge rush of black oily muck and as the valve had to be fully open, it hit the bottom of the oil drum and blew back in your face and all down the front of your boilersuit. I tried covering myself with rags, but the sludge always got through.

Our boilersuits required washing at the end of each watch, we had a washing machine in our washroom but it was out of bounds for boilersuits. We used a 45 gallon oil drum which was kept down on the plates at the front of the engine. A pipe fed steam from the boiler feed pump into the drum that contained a lethal concoction of black oily water, Tepol, soft soap and soda and quite often at the end of the voyage, a bottle of whisky or rum. These would be buried deep in the glutinous morass at the bottom.

It took a very zealous customs man to delve into its murky depths and it always seemed to contain one or two disowned boilersuits rotting away.

The best method for cleaning them was to tow them behind the ship for a night. However there were certain disadvantages, it required a long length of rope; this had to be obtained from the bosun's stores; not easy when you had a Bosun like Frank Russell, a true 'Shellback' if ever there was one.

Frank first went to sea on the Thames Spritsail Barges and crewed on the twelve meter racing yachts of Thomas Lipton the tea magnet. He joined the Merchant Navy in 1936 and in 1942 he was on Union Castle's 'Rochester Castle'. Coincidently she was built by Harland & Wolff and had the same type of engine as the *Rakaia*. She was part of the convoy of fourteen merchant ships, code named 'Operation Pedestal' that set out to get vital supplies to the island of Malta. It was the most heavily attacked convoy of the war and nine of the ships were sunk, despite being hit in number three hold by two torpedoes she managed to keep afloat and was the first ship to arrive.

Frank was one of several men aboard the *Rakaia* who had survived the war at sea, but if broached on the subject he just smiled and said.

"It had its moments where it got a bit iffy."

Another one was Arthur Newlyn the ships Carpenter, (Chippy), to someone meeting him ashore he would have appeared to be an unassuming old chap. No one would ever guess that he had been awarded the British Empire Medal. Unlike Knighthoods, these were not given lightly!

Frank Russell (Bosun).
(Photo Paul Wood)

While on the M.V. *Durham* he went over the side and cut loose a mine with a cold chisel and hammer that was caught up in her paravane endangering the ship and all the crew.

There was also our Chief John (Jock) Cowper, he had been torpedoed twice, but he never spoke about it.

Several of our greasers, together with Gerry the Head Cook had been through bad experiences during their war years. Then there was 'Lampie' the ships Lamp trimmer . Like all Lamp Trimmers that I came across while serving my time, he was a very elusive character, I never did find out where his cabin was. As electricity had superseded paraffin, his lamp trimming days were over, so he effectively became the Bosuns Mate, in charge of the paint locker.

Another problem regarding our boilersuit washing operations was the Captains weekly inspections. If by chance we still had a line out over the stern during one of these, he would pull out a penknife and cut it free, leaving the boilersuit to drift off in the ships wake. Sometimes when we hauled in the line the boilersuit would be missing, probably taken by a shark. We had to weigh the boilersuits down by putting a heavy piece of metal in them, usually a scrap valve or bottom end bolt from the generators, so there were probably a few sharks swimming around that were well ballasted.

In the tropics I found it better to wear white ducks and vest when on watch, you got covered in black oil but they were cooler and easier to wash. It meant you got dirtier but I found it easier to get most of the muck off using a rag and bucket of gas oil before going topside for a shower.

For exceptionally dirty jobs, such as checking the scavenge trunking, we had at our disposal a rubber suit with a hood, so that it completely enclosed the body except for hands. Once worn it was never put on again, it was of huge proportions and required the assistance of another person to help put on; It was more a case of in than on, as once encapsulated in the hideous thing, the sweat just poured from body making the inside so slippery it was impossible to walk let alone work. It was far better to put up with the sludge and wash down with gas oil.

The regular duties of a junior engineer were, to look after and maintain the fuel supply from the double bottom and deep tanks to the settling tank which had to be regularly drained of accumulated water and sludge. Stopping and stripping the fuel and lubricating oil purifiers, the greaser then cleaned them and left them for the engineer to carefully box up and put back on line. The filters on the *Rakaia* were 'Sharples', these were of the vertical spinning type, if they were slightly out of balance after being boxed up, they would vibrate in an alarming manner as they got up to speed. The stern gland had to be checked for leaks and overheating as did all the propeller shaft bearings. He also had to make sure the daily service fuel tank was kept full, look after the boiler and keep the bilges dry and scavenges clear. There were drip trays under all the fuel valves to catch any fuel that leaked from the bleed valves; this was channelled via a series of pipes down to a waste fuel tank under the plates. This had to be pumped back up to the daily service tank every watch.

The junior also had to write up the engine room log about halfway through the watch, this recorded all the temperatures and pressures, engine revolutions, torsion meter reading and the all important sea water temperature. Any remarks concerning adjustments or abnormal occurrences also had to be written in.

Around the world there are places where the sea temperature can change by 40 to 50 degrees in the course of an hour or so. One such place is on the Pacific freeboard of South America; here the Humboldt Current flows up from the Antarctic and can extend up to 650 miles off shore. A ship voyaging from New Zealand to the Panama Canal would hit this current like a brick wall. The Chief always obtained the necessary information from the Captain, so that we knew to within a few hours when we could expect to encounter it. When the time came the chief made sure we were all prepared.

A reading of the sea water temperature was then taken every fifteen minutes. This was read from a long thermometer that was withdrawn from the main sea water inlet pipe that was beneath the engine room plates. As soon as the temperature showed signs of dropping, the valves from the coolers were adjusted so that no significant change took place in the cylinder jackets. Failure to observe these precautions would almost certainly result in one of more cylinder liners cracking, with disastrous results.

On top of his routine duties the junior might have to assist the senior engineer with any repair work that was nearly always being undertaken.

Down in the engine room there was a flagon of concentrated lime juice and a large jar of salt tablets to be taken as and when required. The lime juice was quite useful, as a small drop in a glass of Foster's made a very welcome drink, especially in the tropics.

During the day the Electricians would be about, they had a heavy workload servicing all the electrical motors and looking after the dynamos of the four generators. They also had to contend with the steering gear, which even then in the early 1960s was considered a museum piece. When we reached the warmer seas they could be found out on deck, servicing the cargo winches in readiness for the cargo handling when we reached Australia. Unlike U.K. ports most foreign ones didn't have dockside cranes.

Chief electrician, Alex Henderson & 2nd Electrician, Davey Lang repairing one of the Cargo winch motors.

7
Scavenge Fire

We were three days out of Avonmouth and heading for the straights of Gibraltar, the main engine had settled in and all the temperatures were steady, the flap valve to the boiler required very little attention and a sense of routine had been established. I had just turned in after having a few beers with the 4th, when at about 2.00 am the engine room emergency alarm sounded in the alleyway.

We never closed our cabin doors while at sea, but there was a curtain that could be pulled over while asleep. This was the first time I had heard the alarm, the noise almost catapulted me out of my bunk.

From my early apprenticeship days I had been taught never to run on a ship, so this instinct kicked in, there was no time to put on a boilersuit, we all found ourselves in the alleyway wearing just our jockey pants, with the exception of the Chief who was in his pyjamas and carpet slippers.

The 2nd opened the engine room door and was immediately consumed in a rush of thick black smoke.

"Scavenge fire," he shouted as he slammed the door shut.

"Get some wet rags and wrap them round your faces" he shouted.

The 4th had already grabbed an arm full and was on his way out of the wash room with a dripping bundle. We then wrapped a piece round our heads to cover our nose and mouth and went into the engine room, slipping into our shoes as we went. Once we got down to the middle gratings things improved, on the way down I noticed dark orange flames through the smoke in the area of the exhaust pistons of 5 and 6 units. The main engine revs were down to about 60rpm. The 3rd and 6th had already shut in the coolers and were gathering some fire extinguishers together.

The Chief who was still in his pyjamas and slippers told me to take a reading of the top pyrometers, the 5th did the same for the bottom ones.

It was soon ascertained that the fire was concentrated around 5 and 6 units, the pyrometers were off the gauge and the ones on 4 and 7 were getting high, the paint on the engine was blistering and starting to catch alight. The 2nd told me to go up to the tops and play an extinguisher on the flames.

Another thing that had been impressed upon me while serving my time was to note before starting work on a ship, where the fire extinguishers were and where the ways up out of the engine room were. Once I reached the tops, even though the visibility was zero I was able to feel my way along the hand rail to where I knew an extinguisher was kept. I was soon joined by the 5th, he had carried his extinguisher up from down below, between us we soon had the flames out, but thick smoke continued to pour out of the scavenge casings. While we were dealing with the flames the 2nd cut off the fuel to the offending combustion chambers.

This was done with a heavy quadrant shaped piece of ironmongery that was lifted into position in front of the appropriate fuel pump, once located it would clank up and down until gotten under control by the engineer. It was then lifted by means of a cross bar, so that the fuel pump plunger was lifted clear of the camshaft. Once clear an anchor pin was inserted to keep the plunger up, making the pump inoperative. Once the fuel supply was cut off the temperature would start to come down.

In theory the main engine should be stopped until the fire had burnt itself out, in practice this was not an option. It could take hours before it was thought safe to restart and compressed air being blown through a burning scavenge trunking would almost certainly cause an explosion.

The only way to deal with a scavenge fire was to cut out the hot units and reduce the revs, usually if one was cut out another would catch fire. If the revs dropped to the point of stalling (about 25rpm) the unit with the lowest reading would be cut back in. This procedure was kept up, with suitable applications of foam until the fire burnt itself out and we managed to get all cylinders working again. Once this was achieved the revs would be slowly brought up to normal running speed.

During a scavenge fire, huge clouds of dense black smoke would be pouring from the funnel. If there was a following wind it enveloped the bridge and gave the deck mob a taste of things down below, it also gave the Cadets plenty of work cleaning the smelly residue off the superstructure.

Occasionally with a serious fire, all the extinguishers would be used up, this kept the Donkey man busy refilling them out on deck, to be relayed back down below. In these circumstances all the paint would be burnt off the upper parts of the engine and the whole top of the engine room casing would be given a thick coating of oily soot, giving it a grim and forbidding look. It would require a couple of days soogeing* by the engine room labourers to get it cleaned up.

Scavenge fire.

*Soogee was a concoction of various ingredients such as Caustic Soda, Teepol and soft soap.

By the time the fire had burnt itself out and the engine was back to its normal revs, it was 6.00 am, there was just time to get cleaned up, have a couple of beers and get some breakfast before going on watch again at 8.00 am.

I soon discovered that scavenge fires were almost routine on the *Rakaia*, as we learned to cope with them; there became no need to raise the alarm, unless a particularly severe one occurred. If one looked like becoming serious the engine room telephoned the Chief, who then used his judgement on who he sent down to assist.

This meant that if the alarm did ring in the engineer's alleyway, there was a serious situation occurring down below. One such occasion happened just after passing through the Straights of Gibraltar. The alarm went off at 3.00 am, again the 3rds watch; this was known as the 'Graveyard Watch', because at that time of night almost all hands were in their bunks, the lack of activity seemed to make the ship feel quiet and deserted.

All hands rushed down the engine room to find the main engine slowing down, before we reached the main control platform it stopped altogether, the first time since leaving Avonmouth. The 3rd had already shut in the coolers, so apart from being adrift in a busy shipping lane everything was under control.

"Och she's gassed up." exclaimed the Chief.

This was a symptom not uncommon with double acting Burmiester & Wain engines. The cause was due to combustion pressure entering the fuel system through an intermittent fault with the needle valve of a main fuel valve. This was why a regular eye had to be kept on the exhaust pyrometers temperatures. If a drop in temperature was noticed, opening the bleed valve would usually cure the problem.

On the *Rakaia* the engineer had to walk along the lower grating and check each unit in turn, then climb up to the top grating and do the same. By the time he got to the last one, any one of the previous ones could be playing up. With all the other things that had to be attended to, it's easy to see why a faulty one could go unnoticed before giving trouble.

To get under way again we had to disengage the fuel circulating pump from its driving rod and operate the pump by hand using a long bar, then all the thirty two fuel valves had to be bled of air. Once the engine was primed and ready to start, the bridge was informed. The Chief always took the controls on these occasions and insisted that every one except 2nd and 3rd vacated the engine room either into the shaft tunnel or to the main deck emergency exit. The 2nd and 3rd would remain to man the indicator cocks. This was because a couple of years earlier, a Union Castle line ship, the *Capetown Castle* had stopped under similar circumstances. When the engine was re started a fireball engulfed the engine room killing six engineers and a greaser.

The cause was traced to a small quantity of oil in the air start system that was blown back into the air receiver that then exploded and sent a sheet of flame through the engine room. This was one of the reasons we had to regularly blow down the compressor drains when manoeuvring the main engine.

On this occasion the bridge rang down full ahead on the telegraph, the Chief gave the engine a blow over, the indicator cocks were shut and the engine was started. As usual this was a straight forward process and we were soon under way again, much to the relief of the mob up top.

Once again I had lost a couple of hours sleep, but this was all part of the job and we all accepted it without any hint of complaint. Before the voyage was over, I was to discover that much worse was in store for us.

There is a saying. 'Worse things happen at sea' and from a shore persons point of view there is a lot of truth in it.

It was mid January and we were now making steady progress through the Mediterranean, with around a thousand miles to go before reaching Port Said. After lunch I went out on deck and saw the outline of Malta in the distance, I had always been under the impression that the Med was always nice and warm but it was cold and dismal, just like back home and not the place to linger, so I returned to the welcoming warmth of the accommodation for a noggin and natter (known as swinging the lamp) with the Fiver who told me to make sure I kept my cabin locked once we reached the Canal.

8
Crossing the Line

Two and a half days later we picked up the Pilot and entered the breakwaters of Port Said to tie up to one of the large mooring buoys to wait for orders and pick up the Canal Pilot before proceeding into the Canal. We were immediately surrounded by a large flotilla of small craft, all jostling for position, there were small sailing craft (falukes) with one tatty looking sail, most were rowing boats with all sorts of commodities piled up in them, these were the famous Canal Bumboats.

I was lucky to have been off watch so I was able to see what was going on, this was my first sight of a foreign country, there was not much to see of Port Said but there was a definite atmosphere pervading the ship. A motor launch came alongside bringing the company's agent and the ships mail, we were flying the Royal Mail flag, so I think that gave us a certain amount of preference regarding time schedules.

Within minutes the Bumboats were doing a roaring trade with the crew, when a final deal had been struck with the boatman the money was lowered down to him in a basket and the goods were then put in the basket and hauled up over the ships side. I had been advised by the lads not to buy anything, all of it was rubbish,

The wrist watches were all one winders, by the time you got through the canal they would have stopped, never to go again. The other popular buy was the toy camels; these were stuffed with all manner of filthy materials that guaranteed an infestation of bugs into the accommodation. The prices were extortionate to begin with, but after a lot of bargaining and arguing, particularly at the end of the stop over they could be had for next to nothing.

I was advised that if I wanted to buy anything the best thing to do was to wait until we reached Aden, where the goods were of a slightly better quality.

Besides the Bumboats, there was a whole load of characters that laid out lengths of carpet at prominent positions in the alleyways and out on the hatch covers selling a vast selection of brass ornamental ware, such as pyramids, camels etc.

They also set up their Hubble Bubble apparatus at various places, before squatting down to puff out foul smelling smoke that infiltrated into every corner of the accommodation. These people came aboard supposedly to look after the mooring ropes during the transit of the canal, but it was doubtful if they knew what a rope was, let alone handle one. They always kept secreted about their person a selection of dirty postcards that were slyly shown to you if you stopped to look at their goods.

I had been warned to keep my porthole and deadlight screwed down and my cabin door locked while we had our passengers aboard as they were inclined to take anything they could lay their hands on. I was told of an instance on a previous trip where one of the lads woke up to find all his clothes and movable items had disappeared from his cabin while he was having a couple of hours sleep. Unfortunately he had left his porthole open and thieves had cleared out his cabin using a telescopic pole with a hook on the end.

After these kind of occurrences, retribution came quickly, fire hoses were turned on the traders out on deck and down into the Bumboats. Various overboard discharges were opened so that bilge water and other unmentionable semi liquids were pumped into the boats alongside, causing havoc among their native crews. It was all part of the canal transit and everyone seemed to enjoy it, even some of the locals.

It wasn't long before I experienced the cunning and expertise of the Arabs at first hand. Three or four of us were having a few beers in the 4ths Cabin, when suddenly in walked a Trader wearing a full length Dishdash, we were instantly on our guard. All the time he was trying to sell us some of his junk, we watched his hands like hawks, we eventually worked him out into the alleyway and shut the door. We continued our drinking, at the same time congratulating ourselves on a job well done.

After a while we dispersed to get ready for dinner, I was just locking my door when the 4th burst out of his cabin shouting, "I'll kill the bastard."

It transpired that while we were watching our Arab visitor's hands he had stepped into the fourths shoes under cover of his Dishdash and calmly walked off in them. Unfortunately it was too late for recriminations as we were preparing to get under way. This included fixing a large searchlight on the bows as the last part of the transit would be in darkness. We still had our Arab transit gang on board and the 4th was all for throwing the lot of them over the side along with their trade goods. The Chief advised against it saying it would cause an international incident; the best thing was to inform the rest of the crew and get everyone to boycott them. This would hit them in their pockets as trading with the ships was their only source of income. Most of them worked on a commission, the goods were the property of shore side traders, so our friends would get a slap on the wrist, or perhaps worse, when they reported back to their masters.

The canal transit was made in convoy with several other ships, in most places it wasn't permitted to pass a ship coming in the opposite direction while steaming, one had to tie up. This was the raison d'être our purveyors of ornamental rubbish were on board. I was told our convoy would pull over at Lake Timsah to let a North bound convoy by and to change Pilots, this would be about the half way mark.

I had a couple of hours before going on watch, so I was able to see something of the Canal, I have to say it was a bit bland, the weather being dull and overcast did nothing to improve matters, it was however somewhat warmer than when we left home. After we passed the Canal Company's offices we entered the Canal proper, on our port side was the vast expanse of the Arabian Desert and on the Starboard hand was a large lake. There was not much else to see, just an occasional hut or ragged looking character squatting down on the bank accompanied by a sad looking donkey. The only camels I saw were the stuffed toy ones displayed on the traders carpets on top of number four hatch. We were on six hour watches for the duration of the transit, so it was soon time for me to go below and do my stint.

If all went well it usually took about 12 to 15 hours to get through, this depended on how many times we might have to stop to let another vessel pass. The total length of the canal was about 87 miles. When I arrived down below everything was going well, I found that the boiler had been changed over to oil and was in the care of the Donkey man, so that relieved us of having to keep climbing up to check it out. There was a notice chalked on the board to say that we were on high suction. This meant that the cooling water was being drawn through a strainer just below the water line, if it had been left on low suction we would be sucking in huge quantities of sand from the bottom of the canal and depositing them in the coolers as well as grinding away the interior of the pipes.

We stopped for a few minutes to change Pilots, so we knew we were about half way through, the next significant place of interest would be the crossing of the Great Bitter Lake, about 15 miles ahead. Everything went smoothly for next hour or so, when suddenly a scavenge fire broke out in the rear two cylinders. By now we were all used to dealing with them, but the situation did cause a certain amount of consternation up on the bridge, especially with the Pilot.

The bridge rang down and said there was a bit of a problem, it was now dark up top and as we were the second ship in the convoy, the smoke pouring out of the funnel was blanking out the searchlight of the following ship. Luckily we were entering the Bitter Lake, so we were informed that the ship would divert from the convoy until we got the fire under control.

The only thing to do was to keep going round in circles until it burnt itself out. After the usual cutting in and out of fuel pumps and giving the fire extinguishers a good exercise, we were ready to rejoin the rear of the convoy, leaving the Great Bitter Lake with a cloud of thick oily smoke descending over it to give the locals something to think about the next morning.

We dropped off the Pilot and our friends the Arab traders at Suez and entered the Gulf of Suez. Everything was running well and we were on routine 4 to 8 watches and looking forward to some warm weather.

So far I had not been seasick, but then again we hadn't encountered any rough seas, even the Bay of Biscay had been kind to us. The first day out of Avonmouth the junior said he said felt a bit queasy and was off his food for a while, but after that he was fine. Strangely I never encountered any engineers who suffered from a bad case of Mal De Mer in my blue water days. Perhaps we were too busy to think about it, or maybe the regular intake of oil, smoke, noise and asbestos had something to do with it. Then again it might just have been the beer!

We were soon through the Gulf and into the Red Sea, with three days to go before taking on fuel at Aden. The process of refuelling or bunkering as it used to be known, was the responsibility of the 4th Engineer and his watch keeping assistant the 7th, which was myself. The fuel came aboard at 300 tons an hour, so it usually took about 4 hours to refuel the *Rakaia*.

This was always a procedure that was accompanied by what would be termed in the present climate as a certain amount of stress and today would be deemed to require counselling. The main reason for this was the worry about spillage. As the fuel came aboard it had to be distributed to the various fuel tanks, some were known as double bottom tanks, some were deep tanks. Then there were the Port and Starboard aft tanks.

I had learnt what valves did what and the order the tanks had to be filled without having to stop the fuel being pumped aboard. There were gauges down in the engine room that the 4th kept a close eye on, these indicated what was in the double bottom tanks, the deep tanks had to be sounded with a metal tape. Sounding the deep tanks required leaving the engine room and going into places otherwise never visited.

Consequently it required a lot of running about for the 7th; and as all the tanks began to fill, certain ones had to be shut off, diverting the flow to other tanks as necessary. When a tank that could not be gauged from the engine room was nearly full, a certain amount of judgement had to be carried out by the 7th on the time it was going to take to get back down the engine room to shut the valve before the tank overflowed. We therefore used a system of signals, by banging on the appropriate pipe with a wheel key, which we always carried; the sound would transfer into the engine room, enabling the 4th to shut the valve.

The biggest problem was with the after tanks, no matter how slowly they were filled, they always developed an air lock quite often before they were even three quarters full. This usually manifested itself in a deep rumbling sound, like distant thunder, that was followed by a vibration that could be felt down in the engine room. It was a sort of early warning system that came too late, despite shutting the valves; it was a prelude to a large spillage of fuel oil from the vent pipe on to the after deck.

Such was the predictability of this occurrence the Chippy always built a foot high cofferdam around the vent out of sand bags, it also gave the off duty crew members an excuse to break open a case or two of beer while they waited from safe observation points for the inevitable to happen.

Just like at Port Said we were surrounded by Bumboats, unfortunately I missed all the excitement, but I understood from what I was told later that Chippy's cofferdam sprang a leak, which poured overboard out of the scupper. The crew of the Bumboat under our counter received a good shower and their boat ended up with quite a few gallons of free diesel oil in its bilge.

Before we got under way I bought a very nice transistor radio from one of the fierce looking characters who worked on the fuel station, he came aboard to make sure the main fuel hose was connected ok. It cost me £10 and I had to get a sub off the Chief Steward to pay for it. At the time it was a state of the art bit of technology from Japan, examples like it were only just appearing in the U.K. and I was told it would fetch £50 in Australia so I would be on a little earner. The lads had given me good advice when they told me not to buy any thing at Port Said, the quality of goods were much better in Aden.

Once refuelling was completed we soon got under way, we had been on duty for 8 hours and were pleased when we were relieved by the 3rd and the sixer. That evening a notice went on the officer's notice board to say that Whites were to be worn from 0600 hrs. The first time I put on my white outfit I felt very self conscious, but when I joined the other lads in the Saloon for breakfast it all seemed perfectly natural.

It was a brilliant hot sunny morning with a flat calm sea, all the long hours down below were forgotten as I made my way down aft to the top of the spud locker for a morning of Bronzing (sunbathing). It didn't take many days to get an all over tan which we hoped would put us in the running with the girls in Australia. The sea temperature had increased to a point which made it necessary put our second lubricating oil pump on line, with both pumps working hard, sufficient oil pressure was only just maintained.

We made steady progress across the Indian Ocean, averaging 14.5 knots, the routine occasionally broken by the odd scavenge fire and more frequently with trouble with one or other of the four generators.

This was mainly bottom end bearings; they were prone to running their white metal for no obvious reason. I had a certain amount of experience of scraping and fitting brass and white metal bearings while apprenticed to The London Graving Dock Company so I was able to tackle these jobs with confidence. This might well have been the reason why the old Chief had promoted me from junior to 7th on my first day. He would have known that I had worked in a ship repair yard and that I would be conversant with the black art of fitting white metal bearings. If I had remained as the junior I would have been put with the freezers as soon as the ship started loading the frozen cargo, effectively becoming the 3rd Freezer until we discharged.

Every Saturday evening there was a film show out on deck, the screen was erected in front of the aft mast and the projector was positioned at the after end of the boat deck. It was all operated by the Chief and 2nd Sparks (radio officers). Unfortunately being the 7th Engineer, I had to be on watch during the show; however it was repeated on Sunday afternoons in the officers smoke room for crew members who couldn't make it on the Saturday.

One morning soon after leaving Aden I couldn't help noticing that Chippy was busy on the after deck with an assortment of hatch boards, he had a group of cadets with him and I could see that he was giving them some sort of instructions. I was curious to know what was going on, but time was short as I had to get my breakfast before turning to. Towards the end of my watch a Cadet rang down and asked for more deck pressure, this was a regular thing early in the mornings when they were hosing down the decks and (accidentally of course) those of us who sometimes slept out on deck, but it was an unusual request just before midday.

After lunch I went out on deck as usual to soak up some sun and discovered what Chippy and his gang had been up to earlier on. Down aft, what at first glance appeared to be large box like object soon showed itself to be a swimming pool.

It is impossible to convey the extreme pleasure we experienced, leaving a noisy engine room after four hours in a temperature of around 140 degrees to partake of a dip in Chippy's pool with a can of ice cold lager in the middle of a flat calm, irridecsant blue Indian Ocean.

Money couldn't buy such a luxury!

Cadets Sports Day, On the right Chief Engineer, Jock Cowper umpiring & on the left Pete, the PTI, keeping score. Note the cinema screen on the mast!
(Photo Paul Wood).

However there was a darker side for the pools existence, we were getting close to the equator, which meant that we could expect a visit by King Neptune and his gang of courtiers as we crossed the line. All first trippers or Pollywogs as we were known were bought before him to account for their crimes against the sea; once we had crossed the line we were given the title of Shellbacks.

As I was one of the first trippers I was getting a bit apprehensive about the outcome as I had been told that King Neptune didn't take prisoners.

Another Pollywog was the ships Surgeon better known as the Doc, he was in his forties and was a short, stout, individual with black hair and was a great character. Being Irish he liked his Guinness and was a frequent guest down in the engineer's accommodation.

It was rumoured that he had bought his qualifications in Ireland; this was often substantiated by some of the treatments he prescribed up in his sick bay. However there was probably a method in his practice as there were very few crew members who dared to venture there. Statistically we were probably the healthiest ship in the Merchant Navy.

There was an occasion when the fiver developed a chronic toothache but he couldn't pluck up the courage to visit him. During one of our hydraulic lunches in the Chief Freezers cabin we were joined by the Doc who eventually persuaded the Fiver to let him have a look.

Using the Chief Freezers torch he had a good look in.

"Ah! I see what the trouble is, I'll be back in a jiff." He said as he disappeared into the alleyway.

He soon returned with handful of nasty looking dentist type tools, after giving them

Swimming Pool. (Photo John Layte).

a good swish round in his glass of Guinness he had a good poke about in the Fivers mouth. Then from his pocket he produced what looked like a roll of Elastoplasts, after cutting off a piece he put it over the offending tooth. He stood back and with a smile said.

"That should fix the blighter."

Strangely by the afternoon, the toothache disappeared along with the plaster.

The evening meal that was served at 5 pm in the officers dining saloon was a very formal affair. The Doctor sat at the centre table opposite the Captain; the Doc was very naïve as to the ways of ships and the sea and the old man took every opportunity to exploit it. During the meal the Doc left himself wide open.

"I understand Captain that we shall soon be crossing the equator."

"That's perfectly correct Doctor," boomed the Captain so that he knew he had captured every ones attention.

"Tell me Captain, how exactly can you tell when we cross?"

The old man couldn't believe his luck! and replied,

"There are two ways Doctor, the best one is when we get to the sign post, one side points to the U.K. and the opposite side points down hill to Australia. Unfortunately we are due to pass the post during the hours of darkness so it is unlikely we shall see it. However the second method is just as accurate, you have probably noticed that the water in your wash hand basin drains away in a clockwise direction. The instant we are on the equator, the water goes straight down. Once we are over, it goes in an anti-clockwise direction; we then know that we are in the Southern Hemisphere."

"Fascinating, absolutely fascinating, replied the Doctor as he tucked into his pudding of 'Bomba Alaska'."

It was impossible to keep a straight face; everyone made a dive for their table napkins.

The old man followed this up by telling the Doc that he was expecting King Neptune to come aboard at nine o'clock in the morning and to keep a low profile as he might well be summoned before his Court.

This was good news to me as I would be on watch from 8 till 12, so I might get out of being greased and shaved, or worse, before being dumped in Chippy's pool.

The next morning King Neptune appeared on the after deck surrounded by his acolytes, his throne had been set up and from it he read out from a scroll, the names of the 1st trippers that were to be bought before him. Their punishment would depend on what crimes they had committed against the sea.

Luckily I was on watch, so I missed it all, but the Fiver told me all about it over lunch. The crimes included throwing cigarette ends overboard, throwing empty beer cans and bottles over the side, apparently certain tracts of the seabed were turning into glass highways, especially between Panama and New Zealand.

The engineers in general were accused of leading the good Doctor into bad habits, but as all of them except the junior and myself were Shellbacks they were deemed to be above the law. As I was down well below the water line on watch, I was given a dispensation provided I paid a fine of one case

King Neptune & his acolytes preparing to hold court. (Photo Paul Wood).

of beer that was to be left at the after end of the boat deck, no later than 1400hrs.

The junior incurred the full wrath of King Neptune and was dealt with accordingly. The Doctor was the last to be summoned before the Court and when he didn't appear the courtiers were sent to fetch him from his cabin. He was blindfolded and carried screaming down to kneel before King Neptune where his crimes were read out.

He was found guilty of severely depleting the ships stock of Guinness and associating with the engineers and was sentenced to be greased, shaved and ducked in the pool. One of the things that we always thought a bit strange about him was that he always kept his shirt on when we were on deck, while everyone else only wore a pair of shorts. Once King Neptune's courtiers had stripped him it became obvious to all. He was covered in long black hair from head to toe.

This gave Neptune's Barber a whole new perspective to work to. Instead of shaving him with the large wooden razor, a pair of clippers was somehow produced and he was given a set of stripes that went up his back, over his shoulders and down his chest and then dumped in the pool. The end result gave him the appearance of a Hyena. The cause of the Docs life long embarrassment had been rectified! From then on he was quite happy to sit about on deck displaying his new image.

A few days later a notice went up on the notice board which was just outside the officers smoke room. Water rationing will begin as from Midnight. It went on to say water would only be available at given times in cabins and washrooms. This didn't give any real inconvenience to the engineers, as we were the ones who were in control of the fresh water pump down in the engine room. It wasn't so good for the Cadets as they were always using the showers.

This was a new experience for the crew, including the old Shellbacks like the Bosun and the Chippy. Even the Chief said he had never seen anything like it so soon after leaving home. The Chippy whose job it was to sound the fresh water tanks every day was adamant that there was a serious leak. However the general consensus of opinion was that it was all down to the Doctor. It got round the ships grapevine that he had kept the taps running in his wash basin for hours waiting for the water going down the plughole to change direction so that he could tell the Captain the exact time that we crossed the equator.

9
Through the Tropics

We had been at sea for nearly a month and everyone had settled into the ships routine. The weather was perfect; the sea was the colour of rich amethyst and absolutely smooth. As the ships bows cut through the sea, dozens of flying fish came out of the water like projectiles and skimmed across the surface in all directions. Later when in rough weather we often found them on deck in the mornings, some were about the size of large Herrings. We were also accompanied by Dolphins that played around the bows, rolling on their sides to seemingly peer up at us with a friendly smile. In contrast in our wake some distance astern the occasional sinister black fin could be seen breaking the surface. The waste from the Galley kept these unwelcome guests lurking a hundred yards astern.

Providing the machinery behaved itself, the voyage down through the tropics was an experience not to be missed. Under the eye of the Bosun the Cadets rigged up nets on each side of the after part of the ship.

The *Rakaia* was unusual as far as cargo ships were concerned, in that her after decks were laid of wood which enabled games such as cricket to be played. She was built as a passenger ship for the Ministry of War Transport and was launched in 1945 as the *Empire Abercorn*.

I would suspect that her layout played a large part in her being bought by the New Zealand Shipping Company for conversion into a Cadet Training vessel.

Cricket matches were regularly held between different departments, the wooden deck made for a very fast ball and it would come off the deck like a bullet. The runs were very low; anything over six would give the batsman a celebrity status. This could only be obtained by managing to get one run, then having the luck to hit the ball over the accommodation, past the funnel into the sea, this would be six and out, a feat very rarely accomplished.

Providing the weather was fine, Saturday afternoons were a time for a lot of deck activity for the Cadets. The after wooden decks were perfect for their recreational athletic pursuits, these were overseen by the physical training instructor (PTI). They also had a seamanship instructor together with Frank Russell the Bosun to look after their other deck activities.

It was possible for everyone to include themselves in these pastimes, but after a four hour stint down in the engine room the engineers preferred to sit around and drink ice cold beer which we collected from the brine room on our way up out of the engine room. It was known as the brine room because it was where the fridge departments Brine Pumps were located. It was accessed by opening one of a series of small apertures that had a door with cork insulation about a foot thick on the back.

The temperature inside was well below freezing and all the pumps had a thick coating of frost encasing them. It was a weird experience removing a case of beer from inside, as the aperture was only just big enough to crawl through, when you reached in to lift out the beer, the sweat on the top half of your body began to freeze, while the bottom half was being cooked in about 140 degrees. By the time you crawled out backwards dragging the case (24 cans) your eyelashes would

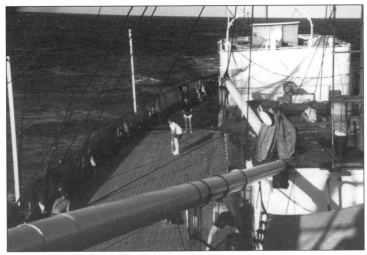

Deck cricket. (Photo John Layte).

72

have a coating of ice on them. A case of Lager would freeze solid in no time at all, if the top half of one of the cans was cut off, a nice refreshing Lager ice lolly could be enjoyed while watching the Cadets do their stuff out on deck. Guinness on the other hand, never froze solid, it only turned into an icy sludge that had to be consumed with the aid of a spoon.

With the onset of hot weather the ships complement of cockroaches became more active, this heralded the start of the cockroach racing season. The race track was prepared by spraying lanes of insect repellent from aerosol cans that we had purchased from the ships Slop Chest.

The cockroaches that had previously been caught and painted in owners colours were placed between the lines under starters orders. When the starting flag went down the roaches were encouraged to move forward by spraying from behind.

Usually they only needed a squirt to get them scampering between the tracks to the winning post ten feet ahead. There were no rules involved, but a certain amount of skill was required as to how the spray was regulated and applied. Too wide a spray could disorient the insects and slow them down or even kill them. Some moved at incredible speed while others ran into the sprayed lines and became confused and some turned back towards the start adding to the general chaos. Losers were executed on the spot while hopefuls were given a second chance.

Cigarettes and cans of beer were the betting mediums; a champion cockroach could be quite lucrative. The engine room greasers always had some of the best, their accommodation which was underneath the engineers seemed to breed the biggest, probably because it received more heat from the engine room.

In those days we had no such luxury as air conditioning, the best that could be hoped for was a galvanised cowl that was poked out of the porthole to funnel in a breeze as the ship made headway. Chippy's swimming pool went a long way in helping the cooling down process, but for me it had a downside. Going on watch following a dip in it, I had a funny sensation of being in another place, the main engine seemed like it was on silent running and I was having trouble orientating to my surroundings. I suddenly realised that I had gone deaf.

What followed was probably one of the worst four hours that I had ever spent. There was a saying that racing car drivers drove by the seat of their pants, in many ways this was so for marine engineers. Subconsciously we tended our engines by sight, sound, scent and feel. Quite unlike engineers of today, who sit in air conditioned control rooms and watch computer screens. To lose one or more of our natural senses put us at a serious disadvantage.

The outcome of this was that I had to pluck up courage and pay a visit to the Doc. The first thing he did when I went into his surgery was to open a couple of cans of Guinness.

"All part of the cure," he stated.

After peering into my ears with his torch he came to the conclusion that they needed a good syringing and said,

"It's that bloody pool you shouldn't go near it, I never do."

He was still getting over the effects of his dunking by King Neptune and his acolytes, so I suppose he had a point.

After blasting some of the Indian Ocean and a bowl full of detritus made up of carbon, gas and lubricating oil out of my ears we had another couple of cans of Guinness and he pronounced me fit for work. After this I have to admit that his prowess as a Surgeon, which was his official rank, rose to great heights as far as I was concerned. My hearing had improved so much that I am sure I could have heard a split pin drop in the crankcase.

As the nights became hotter, we often took our mattresses out on deck and slept there, however it was a good idea to sleep with a pair of socks on as the cockroaches were partial to the hard skin on the soles of your feet. The downside of sleeping out on deck was that you really needed to be up and off the deck by about 6.30 am; otherwise you got a good soaking when the Cadets washed down the decks, which was O.K. but it also soaked your mattress.

From talking to the lads I discovered that a job on the *Rakaia* was not a personal choice among engineers in the Company, she had a reputation as a hard worker and given a choice they would take the better option.

There were two main reasons for this; one was the continual trouble with the generators. They were 'Harlandics' and as the name implies they were made by Harland & Wolff. It was the general opinion that they were probably made for a shore side installation, this was because they gave very little trouble when in port or when very smooth seas prevailed. Once the sea picked up and the ship started rolling about a bit, they tended to give trouble with their bottom ends. As the *Rakaia* was under construction during 1944 there could well have been a shortage of purpose built generators.

For those that like to record these things they developed 335 BHP at 420 RPM; they were rated at 250 KW with 220 Volt direct current. Compared with the screaming, ear splitting, modern day, self destruct generators they were very user friendly. They just trundled round at a speed that in most cases made it possible to detect a problem before it became serious. In those days ear protectors and sound proof, air conditioned control rooms were in the realms of science fiction.

The other main reason for the *Rakaia's* reputation was that four years previously she had a problem with her main engine while on route from New York to Liverpool. She was several hundred miles out into the Atlantic when the main piston rod to number eight unit broke, this caused the con rod, which weighed in the region of two tons to smash through the side of the crankcase and thrash about until the engine was stopped. It was extremely lucky that none of the engineers were seriously injured. The sight and noise of that huge piece of machinery flaying around through the side of the engine must have struck terror into the two engineers on watch.

It's difficult to picture the sheer size of the con rod but the photograph reproduced below shows the Author on the left, holding the jacking spanner that fits the piston rod nut in the same unit that smashed up in 1958. The big lad on the right holding the rope tackle was an engineering Cadet in his last year.

The colossal task of sorting and dismembering all the mangled pieces, isolating number seven and eight units and getting the ship under way was a supreme piece of marine engineering. The day after the disaster the weather deteriorated with gale force winds making working conditions down below very dangerous.

Without going into too much detail, as the episode is well documented elsewhere, the mate and his team rigged up steadying sails made from hatch covers. This reduced the rolling and enabled the engineers to complete the repairs, by which time the ship had drifted a hundred miles off course

The *Rakaia* was able to resume her journey in winds gusting up to force nine, eventually reaching Liverpool with her cargo intact, at an average speed of 6.9 knots with the engine running at 50rpm and assisted by her sails which helped take some of the strain off the main engine. After discharging she went back to her builders Harland & Wolff where she spent three months having the engine repaired.

Probably because of the fairly recent maker's extensive repairs I never thought her reputation was justified, I did six voyages to the other side of the world in her and never experienced a serious breakdown on any of them. It is interesting to note that *Rakaia's* engine was one of the first built to the design of Cuthbert Coulson Pounder, Harland & Wolff's Chief Technical Engineer. In 1944 he was given the official order to build six engines to his design that were capable of developing 8000 bhp.

The first was installed in Lamport & Holt's M.V. *Devis* and two more were installed in Shaw Savill's M.V. *Waiwera*. As the *Rakaia* (*Empire Abercorn*) was under construction in 1944 it is quite possible she was fitted with one of the other three.

The generators on the other hand were another story; they gave trouble on a regular basis and were probably the main reason for her unpopularity among the Company's engineers who had never sailed on her. Despite, or perhaps because of the hard conditions down below she was always a happy ship and tended to keep a regular nucleus of engineers and crew. An example of this was Frank Russell the Bosun and Chippy Newlyn, they had been on the *Rakaia* for dozens of voyages.

Any engineers who were not au-fait at fitting bottom end bearings were soon indoctrinated into the black art of scraping and bedding in white metal shells so that by the end of the voyage they could be considered experts. The generator crankshaft crankpins were 8.5 inches in diameter, but over the years they had been ground down to a variety of different sizes. When one ran its white metal, reference was made to the log book to check the diameter and a suitable pair of shells was selected from the stock that were bolted to one of the engine room bulkheads. Each one had its bore size painted on it so that I could be quickly identified. The one that was the nearest undersize was selected so that the minimum amount of scraping would be required.

When outward bound it wasn't important to have a generator out of action, but on the homeward run all four would have to be kept running.

This was because power to the fridge department had to be maintained at all costs; otherwise the condition of the frozen cargo would be put at risk.

When carrying apples a change in temperature of more than plus or minus one degree Fahrenheit would ruin the consignment. Butter was allowed plus or minus five degrees.

When we had a serious generator problem we had to go on six hour watches, this enabled an extra engineer to be in the engine room so that repairs could be carried out without disruption to the normal routine. If further troubles occurred, all hands had to turn to until the problem was rectified.

In serious instances power had to be reduced to various departments. As we liked to keep in with the Cook, the Galley was always the last on the list. The Galley was at the end of engineer's alleyway we so we were well situated for sampling some of his little goodies.

As the main engine lubricating and cooling pumps were driven by electric motors we sometimes had to slow these down, so that the remaining generators could cope. When this was done the main engine also had to be slowed down to stop it overheating. This upset the crowd up on the bridge, so until everything was running normally we were not very popular.

10
Fire Drill

Every Friday at 4.30 pm we had a fire and boat drill; this was a necessary evil as it disrupted our free time, I don't think any of us ever thought that we would need to take to the life boats in earnest. That is with the probable exception of the old timers who knew only too well the need to take it seriously.

The drill started with seven long and seven short blasts of the ships whistle, everyone who were not on watch had to collect their life jackets from their cabins and report to their stations. Everyone had to know how to prepare the life boats for launching; in fact we were all encouraged to obtain a life boat ticket. This was quite an extensive course and could be taken under the tuition of 'Schooly' up in the Cadets classroom. Once back in a U.K. port a Board of Trade examiner would test your boat handling skills, both practically and orally. Obtaining your lifeboat ticket meant that you were considered qualified to take charge of a lifeboat should the need arise. The main advantage to us engineers was that with the Captain's permission we were able to take the crash boat out, suitably loaded with hydraulic refreshments and explore the local water front and beaches for talent.

The engineers were quite conversant with fighting fires in the engine room, but accommodation and cargo fires were something different, consequently we took this part of the drill very seriously. As soon as the alarm went off it was the 5th Engineer's job to start the emergency fire pump. These obstinate contraptions were always found down Aft at the top of the tunnel escape. (In no hope situations it was possible for the engineers to vacate the engine room via the propshaft tunnel then up an iron ladder to the deck, this was known as the tunnel escape). The entrance to the tunnel from the engine room could be sealed with a watertight door this was activated from just inside the engine room door by a large hand wheel through a series of articulated rods.

The small single cylinder diesel engine that drove the fire pump was always temperamental in starting and the pump being so far above sea level had to be primed, so it was often some time before we had a flow of water from the hose. We always knew when the 5th had got the engine going by the high cracking noise from the exhaust which seemed to be endemic to all emergency fire pumps. It also made the deck around it vibrate like a girl's best friend!

During my apprenticeship I had overhauled several emergency fire pumps, that when finished had to be passed by the MOT and Lloyds surveyors. It was always a tense moment hoping they would start on the first swing.

Another aspect of the fire drill was 'the asbestos suit'! This was a torturous garment that all the engineers had to have experience with. It was made from heavy asbestos cloth with a fully enclosed helmet with a wire reinforced glass window; it came with huge asbestos gauntlets that were almost impossible to grip with.

Once on, the suit gave complete short term protection from any fire, it also gave you an almost certain chance of contracting asbestosis in later life. When it came to my turn

The 2nd Engineer, 4th Engineer and 5th Engineer on a booze cruise with the ships crash boat.

with the suit, an axe was placed in my hand and a safety line was put round my waist. I then had to make my way down to the bottom of the engine room to look for imaginary survivors. What I was supposed to do if I found any wasn't explained to me. I could hardly get myself down the ladders, let alone carry someone back up.

The glass window soon misted up so it became difficult to see where I was going, the whole exercise was fraught with danger and once down at the engine controls the lads on watch gave me a hard time. They placed burning lumps of oily rag on my shoulders and head and tied the safety line to various hand rails. With the lack of visibility and sense of feel and manoeuvrability I would think it would have been impossible to bring someone out of the engine room. I managed to chop through the safety line and climb back to the deck, realizing why the axe had been put in my hand.

There were occasions when laying off a port waiting for a berth or perhaps the Pilot, the old man might decide to have a man overboard drill. A lifebelt would be thrown over the side and the exercise was to lower the crash boat and retrieve it. It all sounds quite straight forward, but in reality the chances of finding the lifebelt were pretty slim. As soon as man overboard was shouted it was absolutely imperative that at least two crew members never took their eyes off it and always pointed in its direction. In lumpy seas with a current running it was very difficult to keep track of it.

The coxswain of the crash boat had to follow the direction the crew were pointing; in the hope of spotting the belt from his very limited field of vision amongst the waves. Luckily man overboard exercises were not carried out very often; otherwise we might well have run out of lifebelts. I think the main object of it all was to make us all aware of what the consequences of falling overboard might be and to keep us on our toes.

Miraculously there are several cases documented where someone has gone missing from on board and the ship has gone about and followed its reciprocal course and rescued the person.

The days passed quickly and the chart on the notice board which gave the daily run showed that we were nearing the coast of Western Australia.

One morning, before going on watch I went out on deck and could smell land. There was a definite aroma in the air and it took me a few minutes to realise what it was. Perhaps it was the continual smell of gas oil and fumes over the last few weeks that had made my sense of smell pick up something different.

When I came off watch at noon I had my first sight of Australia, land was visible on our Port side. We still had a few days to go; the Great Australian Bight still had to be crossed before we reached Sydney.

The more experienced engineers began to talk about their previous visits and their old girl friends; they also gave advice on the best watering holes. As we ran across the Bight we encountered heavy seas, I noticed that safety lines had been rigged along the decks. However the *Rakaia* took them in her stride with just the occasional big sea curling over her bulwarks.

When we came through Sydney Harbour Heads I was on watch, so I missed the imposing entrance to one of the world's finest harbours. The first sight I had was after we stood by, well inside the harbour while we waited to be cleared by the Port

Health Authority of any contagious deceases. We lined up on the deck and the Port Doctor checked us over, for some reason he only seemed interested in our hands. After we had been given a clean bill of health we motored up under the bridge and round to Pyrmont the commercial dock area, to begin discharging our general cargo.

Typical weather in the Great Australian Bight. (Photo Keith Everitt).

Docking day in Australia or New Zealand was always an exciting event, we had been at sea for five weeks and it was the longest you could go without starting to come home again. Ships on this run were built with the knowledge that their engines had to run continuously for weeks on end in extremes of sea temperatures and conditions.

A classic example of the suitability of ships to certain routes was when the N.Z.S.C. purchased for £705,000, Cunards 13,362 ton *Parthia* and in 1962 renamed her *Remeura*. She was powered by steam turbines and was designed for the North Atlantic run. Despite expensive modifications she was withdrawn from service with the N.Z.S.C. after only three years as being found unsuitable for the blue water voyages to the far side of the world. This was due to a combination of reasons such as extreme sea water temperatures which caused trouble with her engines and high accommodation temperatures and with the beginning of air travel the writing was on the wall for the large passenger ships.

The best part of docking day was the mail from home; it was quite incredible how much some of the lads received. It was also a good time for a certain amount of amusement and even condolences. It was a sort of unwritten law that all 'Dear Johns' were pinned on the notice board for all to read and perhaps share their loss of one or more of their girlfriends from around the world. In some instances it released the recipient of any sense of guilt that he might otherwise have had after the first night ashore.

I was told that Tattlers Hotel was one of the favourite places to start our shore leave. It was in this tavern of ill repute that I experienced my first Australian beer. It came out of a nozzle on the end of a flexible hose pipe into our glasses that were lined up on the bar. These were Middies (small) or Schooners (large), it was ice cold and tasteless; however I soon became acclimatised to it and its effects.

At anchor in Sydney Harbour, awaiting health clearance. (Photo John Layte).

The serious drinking lounge was upstairs and of considerable area with many tables.

They had a strange idea of closing at 4.00 pm, probably to hose the place down and then reopening at 6.00 pm with last orders at 10.00 pm.

After leaving Tattlers at closing time we made our way up to the notorious Kings Cross area.

Being a Londoner I was quite familiar with the clip joints and strip clubs of Soho and the East Ends China Town, but I was unprepared for what awaited us in Kings Cross. Without going into vivid descriptions lets just say, it was disappointing and frustrating to discover that the best looking girls all had Adam's apples.

Somehow, we all managed to get back to the ship in the early hours of the morning. After a couple of hours sleep we had breakfast and turned to at 9.00 am to start stripping out the pistons of number one unit for re-ringing.

Before starting work on the main engine we always engaged the turning gear, this wasn't just to facilitate the repair work, but to make it safe to work on. There was always the possibility that the engine might be shunted round by something bumping into the propeller.

The chief decided which cylinders needed work by examining the indicator cards that had been regularly taken by Big Ron the 3rd engineer.

By using a formula and sets of tables the horse power and compression of each combustion chamber could be worked out. It was part of the 3rds remit to take indicator cards every couple of days or so. As the *Rakaia* had an 8 cylinder double acting engine it meant sixteen cards had to be taken during a watch, in reality many more were taken owing to the failure rate caused by oil and other contaminants being blown on to the card at the critical moment.

Indicators are precision instruments, beautifully made and a pleasure to behold. When not in use they are kept in a green baize lined box that also contains all the tools and adaptors required for their use, including a quantity of the special paper cards on which the diagrams are traced.

They were usually kept and cared for by the 3rd engineer. Time and technology have now made the instrument obsolete, in much the same way as the Sextant for navigating and the Tapley Meter for brake testing and in more recent times the fountain pen for writing letters.

We were due to stay in Sydney for ten days, so overhauling the machinery was our main priority. It wasn't long before the engine room took on the appearance of a scrap yard. There were huge pieces of machinery, some weighing several tons seemingly abandoned in various places around the engine room. A variety of chain blocks hung from conveniently placed shackles and the handrails around the top gratings had 550mm diameter piston rings hanging on them awaiting fitting. The general appearance was akin to a medieval torture chamber.

Inside the crankcase large light clusters had been hung to illuminate the interior to facilitate the work in progress. As the lads dismembered and lifted out the piston assemblies with the overhead crane and transferred them via chain blocks to their servicing positions; the staccato rattling noise from the endless chains dominated the engine room.

Down below we had three generators running to make sure there was enough power for the deck winches. The Australian dockers or wharfies as they were known were and awkward lot, they used any excuse to stop work and walk off the ship. The winches and the cargo derricks had to be exactly to their liking so the brakes and seating had to be adjusted to suit each individual operator. Once they had walked off, delicate negotiations would then take place between the ships agent and the unions. A whole day could be lost before a settlement was reached. This suited us but it cost the Company dearly.

Altogether we had six engine room labourers, three of them were watch keeping greasers and one was the Donkey man. While the repair work was going on the labourers assisted the engineers and cleaned down the engine parts and kept the work areas free from oil.

The work on the main engine was divided among the ports that were on our Australian itinerary, this was to ensure that the work in hand did not interfere with the ships schedules.

During this time we also had to check over each of the four generators before starting the long voyage home.

This would entail fitting reconditioned air start valves, fuel pumps and valves, examining and adjusting the bottom ends and taking a set of crankshaft deflections and adjusting the main bearings accordingly. The lubricating oil was changed and the old oil was put in the dirty oil tank to be purified. The inlet and exhaust valves were re-seated and piston rings were changed if required. When the time came to start the long haul across the Pacific all four generators would be in as good a condition as circumstances would allow.

Now that we were on the coast, the junior and myself were put on night work each doing one week on and one week off. In hindsight this gave us experience and the confidence to deal with any minor problems occurring while on duty. The 2nd set a number of jobs that had to be completed each night, such as overhauling valves or repacking pump stuffing boxes.

Sometimes a generator would play up, so another one would have to be started and the load transferred over and the troublesome one stopped and any remedial work would be started. A cold meal, usually sandwiches and a flask of tea or coffee was left out in a small pantry which was off the officers smoke room. I used to go up and collect it about 3.30 am. As the light in the pantry was switched on a black shadow would instantly sweep across the worktop, this was the pantries resident population of cockroaches. They moved at the speed of light; how they could all disappear without them actually being seen must be one of nature's great mysteries.

The sandwiches were kept safe from the marauders by keeping them in a sealed container. I used to take the tray out on the after deck by the engine room entrance and sit in a chair taken from our mess room and soak up the warm Australian night while enjoying the meal. The only sound would be from the generator rumbling away down below.

One night I was sitting back in the chair with my feet up on the cargo hatch drinking the last of my coffee when I was suddenly confronted by gorgeous mini skirted guest. She screamed when she saw me, as I suppose the last thing she was expecting was to see someone sitting out on deck, covered in black oil having a meal 'Al fresco' at four o'clock in the morning. Unfortunately she was closely followed by one of the engine room labourers who was as surprised to see me as she was. I had literally caught him with his pants down as all guests had to have a pass and be off the ship by 10.30 pm. Theoretically, I could have got him into big trouble, but I just smiled and gave them a knowing nod where upon he ushered her up the companionway towards the gangway. How he got past the Cadet on gangway duty I never did find out.

During day work we had a break ('Smoko') at 10.00 am and 3.00 pm. this was taken in the disgustingly filthy mess room at the top of the engine room. When engaged in heavy repairs we usually had our lunch there as well.

It was a small box like room with no natural light; the only illumination came from a single light bulb that could only be described as a glim.

Every entire surface had a nice coating of black oil which had accumulated from the habitual scavenge fires that plagued the engine room. The table top and bulkheads had an interesting variety of explanatory sketches randomly drawn with a fingertip on their surfaces depicting possible solutions to previous troubles down below. There were a couple of unclaimed boilersuits, so solid with thick black grease that they stood unsupported in one of the corners. On the floor several empty cans of beer drifted about according to the trim of the ship. At lunch times the Steward bought our lunch down to us but understandably refused to enter so we always collected it outside. Sometimes there would be up to ten of us in there and as there were only two chairs we would sit around the perimeter with our backs against the bulkheads and our plates on our laps.

By the time we finished eating the plates would all have a ring of oily finger prints around their edges and the handles of the cutlery would have a matching colour scheme. Before returning to work we left the plates out on deck for the steward to collect, whether he returned them to the pantry or threw the lot over the side I never knew.

For some reason the mess room was never visited by the old man on his Captains inspection tours. Perhaps because it came within the jurisdiction of the engine room, he thought it prudent not to upset the Chief. In hindsight I never saw a Captain enter the engine room, even during a serious breakdown. Although the conditions in the mess room were well beyond the realms of today's health and safety brigade we never had any engineers suffer from a good dose of dirty engine oil and carbon. In fact the laughter and comradeship that was generated from our Smoko's and lunch sessions went a long way in making the *Rakaia* a happy ship for the engineers.

We eventually stayed in Sydney for two weeks; our local watering hole was Monty's a rough seaman's pub where they seemed to think that it was an insult if they didn't invite you to have a fight. I later discovered that this custom was something the Australians had in common with the Glaswegians. Another pub we frequented was The Bunch of C**** but the less said about it the better.

During our stay I managed to visit some of the well known places such as Bondi, Coogee and Manly beaches. Without wishing to offend my Australian friends I have to say that Bondi didn't live up to my expectations. This might have been because it was the only Australian beach that was ever mentioned in the U.K. and was hyped up too much.

Our Donkyman, 'The Oracle' as he was known told me that there wouldn't be any beach there at all if it wasn't for the Merchant Navy, apparently ships arriving in ballast from various parts of the world dumped their cargo's of sand over the side just off Sydney Harbour Heads leaving the current to deposit it on the shore at Bondi. Whether this was true I have no idea.

The other and more logical reason was that I nearly drowned myself there. The surf was running quite high, so in true Pommey style I thought the best thing to do was to swim out beyond the breakers and have a drift about. When

I stopped swimming I looked back and realised that I was in trouble, the rollers were sweeping in towards the beach that seemed to be much further away than I expected. I was obviously in some sort of tide rip; I started to head back but couldn't seem to make any progress. This is when you get that funny feeling in your gut and your legs go weak. I could see one of the life savers special surf boats some way off and thought of attracting their attention to get a lift back in, but instantly dismissed the idea. It would have made their day to have dragged a Pom out of the water and no doubt once ashore they would have given me the full life saving treatment in front of all their groupie Sheila's.

Encouraged by this thought and the fear that the shark alarm siren might go off any minute I somehow found the extra strength to get back into the surf. I was then washed well up the beach, where I lay on my back trying to get my breath back and waiting for my legs to get some life back in them. Once my legs recovered I casually strolled up the beach to rejoin the lads who were busy eyeing up the local talent.

"Where did you go?" Somebody asked.

"Oh I just went for a dip to cool off," I replied.

Manly beach with its back drop of pine trees was a much more user friendly place. We took one of the many frequent ferries across the harbour to a quayside with an interesting marine aquarium close by and then crossed over a strip of land to the beach where a good day was had by all. Our favourite beach was Chainman's Beach, which was in the harbour, it was in a small cove and ideal for barbeques and parties with our new found girl friends.

Several of us went to visit a company engineer who was in hospital on the outskirts of town. I can't remember what was wrong with him but he had been left behind by his ship and was in need of a bit of moral support. We managed to smuggle a few cans of Guinness into his locker and update him with news of his ship. Our visit not only cheered him up, it gave us the opportunity to make contact with the young nurses at the hospital.

During our stay in Sydney I made the mistake of leaving my port light open on one of the rare occasions when I went to lunch in the saloon. On my return I noticed a big black footprint on the top sheet of my bunk, it took me by surprise but I soon realised that someone had entered my cabin through the porthole. Judging by the size of the footprint he must have been a big bloke, how he squeezed through I will never know. Nothing seemed to be disturbed, but I suddenly realised my newly acquired transistor radio was missing from on top of my desk. It upset me to think that someone could stoop so low as to steal personal items from a ship.

It was all part of the game in the Suez Canal zone and was to be expected, but in Sydney I thought they would have more respect for sailor's possessions. When I worked on ships in the London docks as an apprentice, it was an unwritten law that no one ever stole from crew members or from the engine room, as ships tools were sacrosanct. Broaching the cargo was a different matter and was considered fair game. The loss of the radio taught me a lesson and was put down as one of life's learning curves. However it would have been nice to have caught the bastard in the act!

We were nearing the end of our stay in Sydney, all the heavy work had been completed so we were able to return to the saloon for our midday meals. We were all in the wash room getting scrubbed up and as always there was the usual larking about when suddenly in the doorway, screaming with delight and jockeying for the best position were several mini skirted dolly birds. The sight of six or so stark naked bronzed and oily engineers, some with a leg up and a foot in the wash basin showing all their tackle and getting the worst off before going in the showers made their day.

For a second or two time stood still, we couldn't believe our luck! The Fiver made a grab for the nearest one, she had been pushed inside the entrance by her mates but unfortunately he slipped over on a bar of soap before he could catch her. They all screamed and ran off along the alleyway as fast as their stiletto heels would allow chased by a bunch of dripping wet naked engineers. They were followed by the Fourth Mate, who we afterwards found out, had been giving them a tour of the ship. I spoke to him sometime later where he admitted it was the most embarrassing moment of his life.

When in port the floors of the alleyways were covered with a thick covering of brown paper, this was to protect them from the dirty shoes of the shore crowd. By the time we sailed the paper would have hundreds of holes punched through it from stiletto heels, some even piercing the linoleum beneath. An engineer's prowess with the ladies could be gauged by the amount of holes outside his cabin door.

That night was the official ships party, held in the cadets anti room and boat deck, these events were all very officer and gentleman like where we had to wear mess jackets and all the trimmings. The Captain invited local dignitaries and their families; along with a certain amount of local talent, it so happened that the girls from the wash room episode were also there.

Having had a preview, they couldn't wait to dance with the engineers so the Fourth Mate lost out once again.

One of the ships secrets, which was passed on to successive engineers was the existence of a very small hole about two millimetres in diameter that had been drilled through the upper deck head of the wash room. The hole went up to the part of the boat deck that was used for the dance floor during ships parties.

During these occasions the duty engineer connected a high pressure pump to the lower end in the wash room. During the evening the target was selected, this was usually a girl who thought too much of herself. Girls like this were hard to find in Australia and New Zealand, but we were always spoilt for choice in America and the UK.

If the target could be persuaded to dance by one of the engineers she would be guided over the hole. If not we just had to keep her under observation while she danced with one of the Mates, or as was usually the case with a friend. Once she was over the target area a signal was sent down to the wash room by tapping on a vent pipe. The duty engineer then cracked open the valve and a fine but powerful jet of water shot up the dress of the unsuspecting girl. As all the girls wore evening gowns a direct hit was rare, usually the jet was absorbed by the fabric leaving no evidence of the prank.

Occasionally a bull's eye was scored, right up between the legs! This nearly always resulted in the girl screaming and then making a frantic dash for the ladies room. There was a time in New York when the victim after screaming punched her partner in the face.

A group of Junior Engineers with their partners at a ships dance.
(Photo Brian Anderson).

11
Australia

After leaving Sydney we headed north to Brisbane, or to be more precise The Queensland Meat Works which was about fifteen miles down river, where we loaded beef and lamb carcases. We then moved up river to Bothwicks Meat Works to load boneless beef destined to become hamburgers for the eastern freeboard of North America.

These slaughter houses were gruesome places, the smell of death was overpowering. There was no escape from it, at first the only place of refuge was down in the engine room but eventually it penetrated into every corner, it even filled the shaft tunnel with its revolting stench.

No matter how much we washed and scrubbed it was impossible to get away from it. To go into Brisbane we had to order a taxi which we picked up at the main gate, this meant we had to walk through the yard between the sheds. I have to say it was the worst walk I had ever done. Coming back after dark was twice as bad, as the place was running alive with rats.

I also heard that there was an abundance of snakes in the area as they were attracted by the rats. Whether there was any truth in it I don't know as I never saw any. The cattle bellowed away in a compound at one end of the building all night and the next morning they were ushered in and came out the other end in cardboard boxes marked as boneless beef. I have to say that I have never eaten a hamburger since.

One morning as I finished my stint on night work I went out on deck and noticed a group of Cadets trying their luck at fishing. We were some fifteen miles up river from the sea so I was interested to see how they were doing. One of the lads told me that the river was alive with catfish, as soon as the bait hit the water it was taken by one. Suddenly a particularly big one was hooked, after playing it for a few minutes the line went slack, when it was reeled in all that was left of it was the fish's huge head. It had been bitten clean through by something much bigger.

All this caused quite an interest among the early risers, the Donkey man came over and informed the intrepid anglers that the river around the slaughter houses was infested with sharks, they were attracted by the blood and effluent that drained into the river from the meat works, in fact the water around the ship was orange in colour. The Donkey man's tip off somewhat dulled the Cadet's enthusiasm for recreational boating activities around the ship.

Being one of the Company's Cadet training ships the *Rakaia* carried several Whalers of clinker construction. These beautiful little craft were varnished inside and out and were kept in first class condition by the Cadets. They could be used for sail training or rowing where they carried a six man team i.e. five oars and a coxswain so they were very popular for rowing races. The Cadets had developed a very strong team and were practically unbeatable in most of the ports. However they usually met their match when competing against the Maoris or the Fijians as these people took great pride in their rowing traditions.

It has to be said that the engineers never had any inclination for rowing; we preferred something with an engine, such as the crash boat that we could load up with beer and explore the surrounding areas. However due to circumstances beyond our control we were forced into forming a rowing team.

It all started when the Mates had lost by a heavy margin against the Cadets, the Mates had the odds stacked against them because they only had five plus the Doctor to make a rowing team, where as the Cadets had around thirty five to pick from. At dinner after the race the Chief made a remark to the 1st Officer (Mate) to the effect that they should have put in more effort. The Mate who was still coming to terms of their defeat said to the Chief.

"I'd like to see your lot do any better, all they do on their time off is go on the piss."

"Och I'm sure they could give a good account of themselves if they were amind to," replied the Chief.

The result of all this was that next morning there was an official challenge pinned on the notice board from the Mates to the Engineers to a half mile race the following week. This was taken very seriously; the Chief got us all together down in the engine room out of sight of the opposition to plan our campaign. In reality we were in a far stronger position than the Mates, something I suspect they had forgotten to take into account when they issued the challenge.

We had twelve in our alleyway; so excluding the Chief who was in his mid sixties and the duty engineer we had enough to muster two crews. Because of this we were able to race against each other and then choose the five strongest rowers and a Cox to make our team.

As our reputation at the dining table rested heavily on the result, we set about bribing the P.T. Instructor with a case or two of beer and several cartons of cigarettes to give us the benefit of his coaching early every morning for the next six days.

On the morning of the race tensions were running high, but we thought we had a good chance of winning. The rest of the crew thought different and put the Mates down as favourites, this was probably because of our performance while still in sight of the ship. On our coach's advice we made a hash of it and caught plenty of crabs, but once round the bend in the river we got down to serious business.

What they hadn't taken into account was the fact that the engineers, despite their reputation as heavy drinkers were on the whole a pretty fit bunch. By the nature of their work they had to be fit, working for long periods down below in temperatures in places of 140 degrees, sometimes swinging large hammers and manoeuvring heavy pieces of machinery with the aid of chain blocks and muscle power and good food automatically kept us in the peak of fitness. It could be said that we worked hard and played hard.

We certainly did both on the day of the race. The P.T.I. had advised us to let the Mates have the head until the last 200 yards, we had placed two markers on the shore at this point, when they came in line the Cox increased our rowing speed so that we went all out and beat the Mates by nearly two lengths.

Because of the betting, which had the Mates down as favourites the engineers ended up with more beer and cigarettes than even they could reasonably cope with, so it was decided to hold a party in the smoke room and invite the Mates and everyone who had assisted in the race to help reduce our stock. The Mates took their defeat in good spirits and we all had a good time.

The best bit came at breakfast the next morning, the Chief sat down at the centre table grinning like a Cheshire cat, but never said a word. The events of the previous day were not mentioned. This was indeed a total victory for the engineers!

After loading the last of the boneless beef we left Brisbane and cast off for a three day voyage up the coast to Cairns. We had to take on a Pilot as this involved a tricky passage on the inside of the Great Barrier Reef. Just in case any sudden engine movements might be required we also went on six hour watches for the duration. It took a day at sea to rid us of the cloying stench of death that had pervaded every corner of the ship since our arrival at the slaughter houses.

Whaler Racing. (Photo John Layte).

This has to be one of the most interesting sea trips it's possible to experience.

There was always something to look at as we negotiated through a series of uninhabited tropical islands with white sandy beaches. At times we passed so close that you felt that you could almost jump ashore,

96

the sea amongst these islands was a myriad of colours from dark blue, turquoise, dark and pale green and in places foaming white as it tumbled on the exposed reefs.

This was truly a south sea island paradise, or so it seemed. One morning after coming off watch, I was leaning on the Gunnels getting a breath of fresh air before getting scrubbed up; we were approaching a small island that had trees and vegetation coming right down to the waters edge. As we came abreast of it and just as a small cove of pure white sand came into view I was joined by the Donkey man.

"I know exactly what you're thinking 7th," he stated in his all knowing way.

"What's that then Donks," I replied.

"Your thinking that you would like to be on that there lovely looking beach, but I can tell you now, if you was on it you would be wishing you were on this ere boat, believe me, those places are full of biting insects and the beaches are infested with giant land crabs, you wouldn't want to lay about there too long that's for sure and on top of that there's probably no bloody fresh water," the last bit was mumbled, as he had already turned and was walking back towards the engine room.

In due course I found that he was right!

Once again I was down below when we docked at Cairns, so I missed all the lush green scenery that marked the entry to the port. When I came up on deck I found that we were tied up to a jetty and surrounded by mangrove trees. Its one of the disadvantages of working down below, you tend to miss some of the more interesting parts of a voyage such as entering a strange port. Conversely, coming up out of the engine room to find the ship tied up in a strange and sometimes exotic place had a surprising effect, it was as if you had been transported through time.

It was in Cairns that I first met some of the native Aborigines, they were a fearsome looking bunch, but after talking to several I realised that they had a certain sense of humour but at the same time had a look of sadness in their eyes. We stayed for three days and as I was on nights I managed to have a good look around the town and the surrounding area.

It certainly lived up to my expectations of the Australian outback, quite unlike the tourist cattle market that it has become today; it is typical of how the whole world has changed since those halcyon days of the early 1960s.

While we were there the weather was very hot and humid, so a swim seemed like a good idea. I was joined by the 2nd Sparks, but we were advised not to swim in the sea owing to the presence of saltwater crocodiles. The truth of this was borne out as I listened to the local radio station in the Fivers cabin that morning. They announced that a local girl, Mary somebody or other, would be late for school because a large croc was in their garden.

Thinking of my recent Bondi experience I came to the conclusion that it was better to be on the sea rather than in it. With this in mind we decided to visit an open air swimming pool which one of the lads had told us about, it was on the edge of town and only a short walk from the ship.

When we arrived it was deserted and seemed to be out of use, but while we were looking around, a man whom I took to be the attendant or caretaker appeared and said we were welcome to use the pool but apologised for the state that it was in. Apparently a new pool complex was in the final planning stages; therefore this one had been neglected and become run down.

When I saw the greenie brown water I understood what he meant, I didn't really fancy swimming in it. As it was getting very hot we decided to give it a go, Sparks jumped in and started swimming around. For some reason I dived in and almost immediately my hands hit the bottom and before I knew what was happening my head followed. I had obviously dived in at the shallow end, if the pool had been tiled I probably would have been ok, but it had a rough concrete lining which opened up a deep gash just above my hairline. I didn't notice anything at first but after I surfaced and wiped the water from my eyes I saw that my hands were covered in blood.

That was the end of my swimming for the day; we decided that rather than drowning ourselves we would drown our sorrows instead, so we headed for the Pacific Reef Hotel to do just that. The hotel was what I had imagined a classic colonial style hotel would be like and it was patronised by true outback characters.

Our arrival in the bar when they saw the blood running down my face was greeted with all the usual ribald remarks and we were made to feel immediately welcome. When I mentioned that we had been swimming in the pool they all rolled about laughing.

"It looks like one of the crocs in there's had a go at him." One of them said as he studied the gash on my head.

"Christ mate you don't want to go near any water round here, it can be bloody dangerous, just stick to the amber nectar," said another as he slapped me on the back.

They were a great crowd and they gave us a sample of true Australian hospitality for the rest of the afternoon.

We returned to the ship in time for dinner, but as I was having trouble trying to stop the cut on my head from bleeding I decided to give it a miss and get some sleep before turning to at midnight. I should have gone up and seen the Doc as I probably needed a stitch or two, but knowing his unorthodox methods I thought it would be best to keep a low profile.

When I was woken by the duty engineer at eleven thirty the cut had stopped bleeding so I was able to start work without any problems.

We were due to sail for Panama in the afternoon so one of the jobs that the 2nd had put on my list was to put a turn or two of packing in the stern gland.

This is a heavy procedure as the packing used on stern glands is about two inches square and to get it into the stuffing box on the lower part of the shaft meant clambering down the tunnel well and tucking the packing in from underneath. This is a straight forward job when in dry dock but when the ship is afloat and fully loaded the water pressure in the stern tube was considerable. The nuts on the adjusting studs had to be eased back until there was just enough space to extract the first turn of old packing, then two new turns could be inserted. Sometimes this could be done without getting wet, but usually as the nuts were slackened back, seawater would start to squirt in.

As long as the remaining turns of packing didn't get forced inboard, a new turn could be worked in then the whole lot would be tightened up before slackening off again to insert the second turn. It always meant that you were in for a good soaking, so the best thing to do was to strip off everything except your shoes before starting. Even in the tropics the shaft tunnel keeps nice and cool as it's about fifteen feet below the water line and has the frozen cargo over and around it.

By the time you have wrestled a six foot length of two inch square greasy packing into the stuffing box you are not only soaking wet, you are near freezing. After getting the first turn in you had to take a quick walk back up the tunnel to the warmth of the engine room to get warmed up ready for the second turn. Once the job was completed, a hasty return to the engine room to get dried off and a mug of tea was the order of the day. Or in this instance the night. I have to say this was not the image of an engineering officer that most people would have perceived.

We were sailing at 2.00 pm so after breakfast I turned in to get a few hours sleep. We were going on to six hour watches from midday until we cleared the Barrier Reef. As I would not be required down below until 1800hrs it meant I would be able to be on deck as we departed Cairns.

We headed almost straight out to sea on a course that would take us through the Grafton Passage into the Coral Sea; we had a twelve day run ahead of us to Pitcairn Island.

When I went on watch everything was running well, the stern gland was behaving itself with no signs of overheating or leaks. Sometimes when it had been completely repacked during a dry docking it would run hot, usually keeping a fire hose on it for a while solved the problem. Once it had bedded in it very rarely gave any more trouble. Encountering very heavy seas sometimes made it leak, but generally it was maintenance free. Compared to modern stern gear it was very basic in design, but when first developed it represented a giant leap forward in marine engineering.

This came about in 1858 when John Penn & Sons in conjunction with Francis Petit Smith solved the problem of excessive wear and breakages on propeller tail shafts by using Lignum Vitea for the bearings. It's interesting to note that the latter was also the man who patented the first screw propeller.

12
Pitcairn

After a trouble free run, Pitcairn Island appeared like a grey blue cone on the horizon. As we got closer it changed to a verdant green mountain with very steep sides going straight down into the sea.

I was on deck at the time and could make out clusters of white buildings dotted about on the face of the hillside. I was wondering how close we were going to get when I heard the engine room telegraph ring and felt the ship slowing down. We came to a stop about three quarters of a mile from the island and almost immediately I saw two large whalers heading out from the shore and I could see that one had an engine and was towing the other.

They were full of islanders and they were soon alongside and scrambling up the Jacobs ladder that had been put over the gunnels. They were a motley looking bunch and were wearing an assortment of garments and head gear that gave them a piratical appearance as opposed to native south sea

Approaching Pitcairn Island. (Photo John Layte).

islanders, they were all bare footed and seemed to have extra large feet.

Disappointingly I noticed that there were not any females amongst them, but I was told they were not allowed to visit ships any more owing to a bit of hanky panky that occurred during a visit from a passenger vessel. It wasn't long before some intensive trading was taking place. Goods and baskets of fruit were being hauled up the side of the ship and a derrick had been swung out ready to lower a couple of cargo nets full of goods for the islands inhabitants.

The more experienced crew members knew what the best trading items were. A carton of ships soap could be exchanged for a bespoke wooden carving of a flying fish or turtle, or perhaps a selection of tropical fruit.

Nails and razor blades were highly regarded by the islanders; lengths of timber were also highly prized items.

I noticed that Chippy Newlyn had the monopoly for this sector of the trade, there was a steady stream of islanders coming from down aft with a long plank of wood bouncing away on their shoulders. I deduced that these were his regular clientele and had probably put their order in during the *Rakaia's* previous visit. One of the best things about our stop at Pitcairn was the fresh fruit; there was a huge variety, only apples were missing, for some reason they didn't seem to grow them on the island.

Pitcairn Islanders coming aboard.
(Photo John Layte).

The down side of this welcome luxury was an abundance of aggressive insects that lurked the decks and parts of the accommodation for one or two days after we left. I was sitting on number four hatch just outside the engine room later that evening when I felt an intense pain on my ankle; I had been bitten by a small black beetle that gave me a painfully swollen ankle for nearly a week. Why they disappeared after a day or so was a bit of a mystery, the Donkey man's theory was that the intruders were soon exterminated by our resident colony of marine cockroaches.

There were probably thousands of cockroaches on the ship, but they were rarely spotted, they seemed to have adapted to live in close proximity to the crew without being seen, where as these new comers seemed to want to make their presence known as soon as they arrived.

One of the other important things about our stop at Pitcairn was the mail, Pitcairn stamps were a major part of

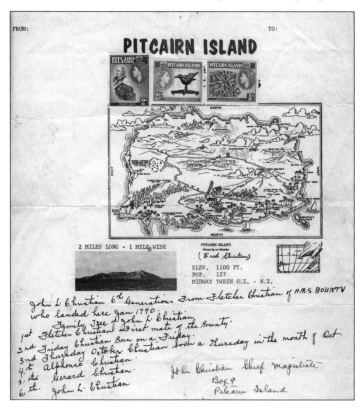

their revenue, most of us bought stamps from them and gave them our letters to post on the next ship to call at the island. In the early sixties visits from ships had dropped to an all time low, N.Z.S.C. passenger ships now took a higher route and stopped at Tahiti instead.

There was a good chance that I would be home before my letter, but as we were going up to Canada it might just beat me. I heard that on average two ships a month stopped at the island, most took a more northerly route when making for Panama.

When all the business was completed the islanders piled into their boats, one of the them started the engine on one of them with a with a handle and as they pulled away, taking up the strain of the towing hawser to the other boat, they started singing the farewell hymn "In the sweet by and by;" it was considered bad luck if they didn't sing as they left.

The Miro wood that was used for their carvings was obtained from Henderson Island which was just over a hundred miles away. To get there they had their boats hoisted aboard one of the visiting cargo vessels and they all hitched a ride. They did this about once a year when the weather was favourable; they then had to sail back to Pitcairn on their own. As Henderson was only inhabited by birds, rats and land crabs, they had to take enough provisions with them for at least a weeks stay.

We passed Henderson Island just before dark; it was fairly low lying and would be difficult to spot in bad weather. 'The Oracle', our Donkey man appeared on deck as we came abeam of the island and gave us a running commentary about some mysterious finds that had been discovered a few years previously.

Apparently a number of skeletons had been found in one of the numerous caves that exist on the island.

"They were probably shipwrecked, fell asleep and got eaten by the rats; the place is running alive with'em, either that or the crabs had'em, them brutes use coconuts for their shells, you wouldn't want to camp out on that there beach, I can tell you." He said this in between sucks, as he flashed up his pipe before strolling back to his cabin amid a cloud of foul smelling smoke.

We had another ten days to go before Panama, the time slipped by quickly with only the usual generator and scavenge fire problems to contend with. Now that we had our frozen cargo aboard, a bit more urgency entered into the generator repairs.

As we neared Panama a close watch was kept on the sea temperature and it was on my watch that it started to drop dramatically, the Fourth phoned the Chief who sent down the junior just in case we needed a third hand to keep an eye on the jacket temperatures.

When I came up on deck at mid day the sea was oily smooth and alive with birds and fish. There were flocks of pelicans diving like guided missiles into dense shoals of fish or squid plus a variety of other sea birds all screeching away as they swirled about over the surface of the sea.

A group of excited Cadets were indicating that there was something of interest ahead of us so I went up to the boat deck to investigate what all the fuss was about. A couple of hundred yards ahead were two huge whales, they had obviously been seen by the First Mate up on the bridge because the ship had already veered off to Starboard to avoid them. Nevertheless we still passed quite close to them and as we did so I heard a loud whooshing sound which sounded just like our main engine being blown over on air and the ship was enveloped in a foul fish smelling spray. One of the whales rolled over and lifted its massive fan like tail out of the water and bought it down with a load smack that could be heard all over the ship. The noise from our propeller must have affected them, but they didn't seem to be at all concerned and they continued to splash about as we left them astern.

The next morning we arrived at Panama, the old hands were busy describing previous shore visits and regaling some of the vivid 'exhibishes' they had witnessed, (some things are best left unsaid!) I was told that if we were lucky we could be through the canal in six hours, but it might be anything up to twelve hours. This was going to be my first transit of the canal, so I was looking forward to the experience.

We went straight into the Miraflores locks, the sheer size of these locks amazed me as the only locks I had seen in the past were in the London Docks. The next surprise was the speed with which we rose up to the next level. After Miraflores we motored a short distance into the Pedro Miguel Lock which lifted us up to the canal. We were now on six hour watches, but as the lock transits were controlled by the canal mules, these are the little loco type tugs that have control of the ship while in the locks; I was able to nip up out of the engine room and see what was happening. The land around the locks and administration buildings was cultivated and manicured to perfection; it was mostly very short grass that would make any golfer green with envy, I was led to understand that it was kept like that to deter the snakes.

Once we cleared the locks we had a straight run to the Gaillard or Culebra Cut, this is a huge cutting of about eight miles long through the continental divide. After passing through the Cut, the jungle came right down to the sides of the canal with an occasional waterfall pouring down from high up on its banks.

I came off watch at midday so I had the rest of the afternoon to take in the scenery; it was very hot and humid with frequent deluges of rain that could best be described

Approaching the Miraflores locks to the Panama Canal.

108

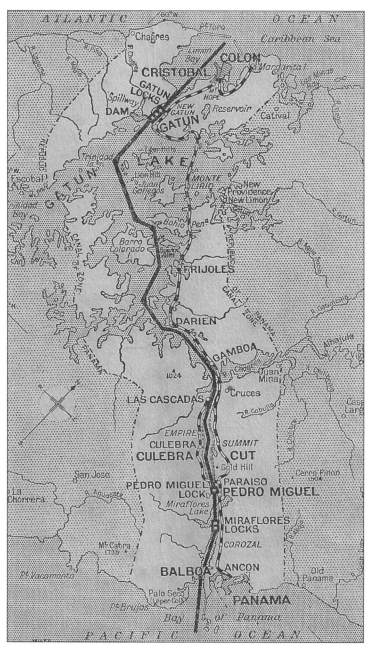

Map of The Panama Canal, note the route of the old railway which ran from Colon on the Atlantic side to Ancon on the Pacific.

as liquid stair rods, the noise as the rain hit the ship made normal conversation impossible. This downpour brought the visibility down to zero so the engine revs were reduced to dead slow.

With no warning the rain would instantly stop and the sun would then come out in force and turn the decks into steaming cauldrons. We then entered Gatun Lake, a vast man made area of water, with small islands in every direction and all covered in dense forest. Rising out of the lake were numerous dead trees showing how the whole area must have been dense jungle before it was flooded. Apparently there used to be plenty of crocodiles around the lake, but by the time of my first visit they had all but disappeared.

The Pilot navigated us through the twists and turns of the channel through the lake and at times we passed very close to one of the dead trees that stood out like white sentinels with just an odd branch still attached and perhaps pointing the way. It made me wonder if the Pilot used the trees as markers to help navigate his route.

As we neared the Atlantic side of the canal I heard the telegraph ring and felt the ship slowing down, we came up to another ship which was at anchor and dropped our anchor a couple of hundred yards astern.

Apparently we were going to have to wait for a convoy coming the other way; no one seemed to know how long this would be. I went back on watch at six in the evening and Stand By was rung down shortly after, so we got under way and headed towards the Gaturn Locks which would drop us down to the Caribbean. We eventually cleared the canal at 8.00 pm that evening so my first transit had taken ten hours.

Ahead of us was a two day run to the Dutch West Indian island of Curacao where we were to take on fuel at the terminal in Caracas Bay. The terminal was over looked by Captain Morgan's Castle, I would have liked to have walked up and had a closer look but once again I was on bunkering duty with the Fourth Engineer. A short distance from our berth was the 'swimming pool', this was a netted off area of beach which was supposed to act as a protection against sharks and was for the use of ships crews.

Close inspection revealed several large holes in the wire net; however sharks never seemed to be a problem. The biggest danger came from stone fish, these creatures are impossible to spot, as they blended in perfectly with the seabed. Despite wearing plimsolls the Chief Lecky was stung on his ankle and remained in a bad way for several days and to make things worse he had to visit the Doc every morning and have the poison drained off. While many of the crew enjoyed themselves ashore I spent the next four hours assisting the Fourth, trying not to have a fuel spillage.

13
Homeward Bound

Once refuelling was completed we got under way, initially our next port was to have been Vera Cruse in the Gulf of Mexico, but for some reason it was changed to Charleston in South Carolina. Once we cleared Cuba we encountered bad weather with fairly rough seas all the way to Charleston, but luckily we had no trouble with the generators. We docked at a wharf on the opposite side of the river to a U.S. Naval Base where many of their ships were mothballed.

As this was our first American port we had to be cleared by the immigration department. We all had our photographs taken and were questioned by a government official about our political views, as this was the height of the cold war I suppose they had their reasons. The interviews were conducted in the officers dining room in a very intimidating manner, when my turn came; the first thing I noticed was that the official was sitting at a table with a revolver in front of him.

I remember him saying that I should be grateful to be granted permission to go ashore in the United States and to make sure that I always carried

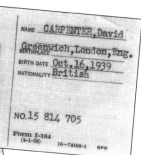

my alien ID with me at all times.

This proved to be good advice as whenever we visited a bar, we were invariably asked for our ID cards. They were very rigid about under age drinking in the States where the minimum age was twenty one. Most of us were in our early twenties, so we soon got used to the routine.

After unloading some of our cargo of boneless beef, we sailed the next day for Philadelphia. I had been looking forward to getting ashore in Charleston but we had to change the piston rings in one of the top exhaust pistons so there was no time.

We stayed in Philadelphia for three days; this gave us plenty of time to fit a reconditioned set of fuel valves to the main engine. They weren't very big but they were quite heavy and required two hands to lift them into position. As there were thirty two of them, the Third would be kept busy overhauling the ones we had removed for some time to come. After they had been fitted we had to prime and bleed the system. We also checked out the main engine internals and hammer tested all the nuts to make sure they rang true.

The best way of describing our runs ashore would be to say they were bland. In fact that was my impression of the whole place. After Australia, Philadelphia seemed very impersonal and aloof, it felt as though there was something lacking. We visited a variety of watering holes during our short stay and none of them had that friendly atmosphere that was predominant in Sydney and Cairns. The most memorable thing about the place was that it was there that I first saw a television program in colour.

The Fiver and I went into a bar and after receiving the usual dour looks from the customers lining the bar counter we went to the far end and there in the corner was a television showing a western film in colour. Back home colour television was still in the realms of science fiction.

After discharging enough boneless beef to keep the hamburger aficionados of Philadelphia happy we sailed for our next port which was New York, or to be precise Newark. This was a run down district with shops fronts covered in wire meshing and displaying posters saying that they had a sale on. The side walks were strewn with all kinds of litter and rubbish, a quick glance up the alleys between the dilapidated buildings revealed where most of it eventually ended up. I was used to London's East End but nothing prepared me for the squalor that pervaded this district of New York. How things have changed, most inner cities in the U.K. now have the same look about them as did the streets of Newark fifty years ago.

The Junior said he wanted to buy a small tape recorder so during some time off we ventured among the back streets or maybe it was the high street, it was a job to tell the difference, until we came across a shop selling all types of electrical goods. Through the iron grills covering the windows he spotted exactly what he wanted, it was priced at $12.

"A bargain," he said.

After he paid for it the salesman told him to go across to the other side of the shop to pay the state tax.

"But I've just paid for it," answered the Junior.

"I know that pal, but you still have to pay the tax," insisted the salesman.

"Well I'm not paying more, it's priced in the window at $12 and that's all you're getting," replied the Junior who was quite a big bloke and perhaps rather intimidating in the way he answered.

It was all starting to get a bit out of hand and while they were at loggerheads I noticed a man pick up a large transistor radio and walk out of the shop with it. I mentioned this to the salesman, who just shrugged his shoulders and said,

"It happens all the time Mac."

He was more concerned about the Junior not paying the state tax than he was about the loss of an expensive radio.

The junior stood his ground and eventually the salesman must have somehow called the police. An archetypal New York Cop suddenly appeared in the doorway, after hearing both sides of the story he solved the problem by saying,

"Well if these guys can prove they are not from these parts they won't have to pay the state tax."

We immediately produced our Alien ID cards and our British Seaman's ID books, where upon the cop ushered us out of the shop saying,

"On yer way guys, but don't try anything like it again."

On our return to the ship we soon found out that the tape recorder wouldn't work! The Chief Lecky took one look at it and laughed,

"You dipsticks have got a lot to learn, you should never buy electrical stuff over here, it's the wrong voltage and runs on different cycles, give it to me, it looks like a piece of junk but I'll see what I can do."

A few days later the Chief returned it in full working order saying,

"That'll cost you a case of beer."

We made several nocturnal visits to Manhattan during our stay; we caught a Greyhound bus which took us right into a skyscraper in the centre of town. From there it was only a short walk to Time Square and Broadway, but I have to say it didn't live up to the images depicted in the Hollywood movies. We had a few beers in Jack Dempsey's bar and moved on to the Great Northern Hotel where the Merchant Navy Officers Club was situated.

This was a brilliant establishment and a credit to New York. We were entertained by the most gorgeous hostesses you would wish to find anywhere and many lasting friendships were made there, There was no hanky panky, dancing to tunes played by Mrs Muir on the piano, was about as far as things went, but we were made to feel welcome in a homely sort of way, what a pity other places didn't do the same sort of thing. It would have saved embarrassing visits to the Doc on the morning after and the queues at the VD clinics would have been drastically reduced in many parts of the world.

During our stay the ship organised a tour of the City on a Greyhound Bus. We visited and had a tour round the United Nations Building, a visit to the top of the Empire State Building and a guided tour of the H.Q. of the U.S. Coast Guard tracking facility.

This was in the days before satellites, so they had devised and made available to international shipping company's a scheme which could keep track of individual vessels in the Mid and Western Atlantic. While in the zone a ship would report its position every twelve hours to the nearest Coast

Guard station; they relayed it to H.Q. who then sent it by undersea cable to the appropriate shipping company.

This way a company could keep track of its vessels while in these waters. It gave a certain amount of safety for the crews, but I suspect it was commercially oriented as it enabled the company to cable their agents and give an accurate time of arrival. The money saved in speeding up docking probably paid for their membership of the scheme.

After New York our next port was Boston, this was an interesting little voyage as we went through the Cap Cod Canal but its downside was that we went on to six hour watches. The canal is only just over seventeen miles long so we were soon through and before long we were manoeuvring into our berth at Boston.

We left Boston after a couple of days and headed up the coast of Canada to St. Johns and then on to Halifax in Nova Scotia. The main engine was running well and surprisingly so were the generators, perhaps it was because we were now in the cold waters of the North Atlantic. The temperature had dropped considerably, this brought up the oil pressure to the bearings so we were now running on only one oil pump. I also exchanged my 'white' ducks for a nice warm boiler suit.

It became too cold to linger about on deck so our leisure activities were now confined to the accommodation. The cockroaches had gone to ground which meant the racing season was now over so reading was probably the most popular pastime. We had quite a good selection of books in the ships library, (this was a bookcase situated in the officers smoke room).

Every now and again we had a swap with another ship so there was always something new to read. The books by Ian Fleming were always popular; this of course was in the days before the James Bond films. I remember when I first saw one some time later, it didn't seem to live up to the book.

We also had darts and chess in the smoke room and some of the lads did a bit of model making. After one of the Cooks powerful curries a blue flame competition might be held. This was a nocturnal event and the best results were obtained with the lights out. There were only two rules; they were that jockey pants had to be worn at all times in case of flashbacks and a bucket of water had to be made available. Some of the results were quite phenomenal.

The next port on our itinerary was Montreal; this was about a thousand miles up the St. Lawrence River. After passing Quebec the temperature began to rise, so much so that a notice went on the board that as from the next morning we would go into whites. It all seemed very strange, one day we were in near freezing conditions and the next in sub tropical ones.

While we were in Montreal the temperature went up in nineties so we were back to white ducks and a heavy workload, culminating in large quantities of hydraulic refreshment in the evenings. This was because the Chief decided to change the piston rings on two of the units prior to our home run across the Atlantic.

Among the many things that were impressed upon me during my apprenticeship was to never trust ships lifting gear, as it was rarely if at all tested. This was especially so with eye bolts and rope tackle, the latter tended to dry out in the heat of engine rooms and the fibres break up. I always gave it all a good check over, making sure the eye bolts had good threads and were not bent. Cracks were a different matter; all we could do was to check for them visually. Luckily we never had an accident due to faulty gear.

We stayed in Montreal for six days unloading some of our frozen cargo and then took on a consignment of newsprint for the Continent. There was a strange atmosphere in Montreal, every one spoke French and seemed to have no great liking for the British. Not far from our berth was China Town, a place we were advised for our own benefit not to visit, which of course we did, but I found it rather surreal to hear a Chinaman speaking French.

We finished our repairs the day before sailing for Hamburg. The loading of the newsprint which looked like giant rolls of brown paper was finished and everything was made ready for sea. The main engine had been checked and double checked then primed and was as ready as it would ever be.

The first start after repair work was always a bit apprehensive. When the telegraph rang for the first movement you could cut the tension with a knife, everyone was secretly crossing their fingers. A starting cock up at this point would be hard to live down, especially in the dining saloon.

For some reason the Mates didn't seem to understand the old Rolls-Royce saying of 'we never breakdown, we only fail to proceed'.

Everything went according to plan, Big Ron the 3rd Engineer calmly operated the controls and before long we were heading back down the St. Lawrence River. The next day we went back into Blues and the temperature began to fall as the sky became overcast, making the landscape of Quebec look very plain and dreary. One of the things that surprised me was the abundance of church steeples that could be seen as we went down the river.

Three days later we dropped the Pilot and headed out of the Gulf of St. Lawrence into a rough and unfriendly looking North Atlantic. The main engine had settled down and was running well; even the generators were behaving themselves. We were back on four hour watches and everything assumed a regular pattern.

As we entered the Western Approaches a strange aura seemed to encompass the ship, the crew began to behave in a manner different from what might be called the norm. Once in the English Channel some people became clearly eccentric. It's a strange feeling, difficult to explain, but to all British seamen it was known as 'The Channels'.

The Chief Freezer seemed to be affected in a different sort of way, he locked himself in his cabin and only appeared when it was time to polish his fridge machinery or at meal times. I had always thought he was a bit of a strange character as he sometimes acted in a way that he must have known would cause comment.

An instance of this happened while we were in Australia; he boldly came into the dining saloon and sat down to lunch wearing a bright Hawaiian type shirt with matching shorts and flip flops. The old man practically had apoplexy; I am sure I saw steam coming from his ears as he stood up and shouted across the saloon.

"Who the bloody hell do you think are you, get out of my saloon immediately."

The Captain was quite right of course, if strict etiquette was not adhered to he would probably have the Chief coming into the saloon in his carpet slippers followed by the rest of his engineers wearing their oily boilersuits. Even to enter the saloon with a badly knotted tie equated to a flogging offence as far as the old man was concerned.

The Chief Freezer kept a pair breeding budgies, or so he called them, in a special nesting box in his cabin where eventually he hoped to have some chicks that he would be able to sell.

Several of the crew had also obtained a variety of parrots during our stay in Cairns, if the weather was good on a Sunday the after deck became an avairian delight. Most birds had their wings clipped so they were able to wander about at will and took great delight in ripping out the caulking on the wooden deck.

I came out on deck just in time to witness a sad little incident; a small brightly coloured parrot ran across the deck and went straight out of one of the scuppers. We all rushed to the side of the ship to see it quickly disappearing astern. The bird fanciers hurriedly gathered their feathered friends and retreated into the accommodation.

A true bird aficionado was Frank the Bosun, he always kept and successfully bred quails. He was always on the lookout for novice bird keepers as they gave him a chance to deploy one of his favourite practical jokes. This trip he had the perfect victim in the form of the Chief Freezer.

Frank connived with Big Ron the 3rd Engineer to secrete a quail's egg in the nesting box of the Chief Freezers potential breeding pair of budgies.

Bets had already been placed that their relationship would produce an egg before the end of the voyage; they had also been placed against a chick ever hatching out.

The Chief Freezer discovered the egg during his morning inspection of the box and posted a notice on the board saying that a chick was on its way. Cases of beer changed hands due to the presence of the egg and betting increased for and against its ever hatching. The Chief consulted his book on birds but it was still some time before he realized that something was not quite right about the egg. It became evident that the budgies were showing no interest in it and that certain amount skulduggery had taken place.

After putting another notice on the board to say that all bets were off due to malpractice, he locked himself in his cabin for the rest of the trip up the Channel. The notice caused a lot of confusion regarding the cases of beer that had already changed hands and been partly consumed. It was a good job we were nearly at the end of the voyage as the cock up would probably take weeks to sort out.

14
Delivering the Goods

We picked up the Channel Pilot and went on to six hour watches which would continue until we reached Hamburg. The English Channel has a reputation for bad weather and has to be treated with respect. Rarely do you get huge seas as in the Atlantic, but short steep seas are the norm with bad visibility or thick fog. These factors and the concentration of shipping made it necessary to have an engineer standing by the main engine controls at all times.

After an uneventful run up the Channel and across the North Sea we docked in Hamburg where unloading started straight away. We only stayed for two days but I did manage to have a run ashore. The thing that impressed me most, besides the Winklestrasse was the water buses. They made full use of the vast amount of water area of Hamburg Docks and were very frequent.

We were soon underway again, this time to Antwerp in Belgium, our last port of call before London. The locks into Antwerp Docks were much larger than the ones in London, easily taking two ships at once. We shared the lock with the *Tabaristan* one of Strick Lines vessels. Again our stay was short, only three days, but together with Jamie and Norman the fifth and sixth engineers I went ashore to sample the delights that Antwerp had to offer. The most prominent being Danny's bar, a haunt known the world over by seafarers. It was a place where it was not advisable to visit the men's toilet, that's for sure!

While on the Continent we only had to do minor maintenance work, the main engine was kept warmed through, ready for a quick departure. We locked out of Antwerp Docks for a short trip across the North Sea to the Thames Estuary. On the way we were notified that our berth in the Royal Albert Dock was full, in fact the whole of the Royal Docks were full.

This meant we would have to a drop anchor off Southend and wait until our berth became available. This was not good news as we had been expecting to pay off the following day. When we arrived off Southend we found that we were not the only ship queuing for a berth, there were at least eight other vessels laying at anchor. One of these was a Harrison line ship; I had worked on many of their ships during my apprenticeship and I knew they used the West India Docks so it looked as though those docks were also full. Halcyon days indeed!

Everything was kept on standby ready for the off which came two days later. It was probably the longest two days of my life; the weather was typically British, cloudy with heavy drizzle for most of the time.

Because of this we were confined to the accommodation and spent our time off trying to find something in the library that we hadn't read before, swinging the lamp about our various shore forages and drinking Fosters Lager that didn't quite taste the same as it did a few weeks previously in the tropics.

We eventually got under way for a short trip up the Thames to the docks and as we passed Tilbury the sun came out so I went out on deck to see what ships were about. I was leaning on the gunnels when I was suddenly enveloped in a cloud of foul smelling tobacco smoke.

"I see they have finished the new jetty, I've berthed at the old one several times in the past when I was on the passenger ships, that one looks like a nice job," remarked 'The Oracle' in his all knowing style.

"It aught to Donks, seeing as I made all the tie rods for it," I replied.

"Christ Seventh, if you don't watch out you'll end up a bigger liar than what I am," he said laughingly as he went back to check on his boiler.

As it happened it was true, it was the worst three weeks of an otherwise most enjoyable five years apprenticeship. I had to screw cut the threads on both ends of over a thousand rods of various diameters. It was done on an old Victorian screw cutting machine that had a mind of its own and was sited just inside the huge doors of the machine shop. It was in the depths of winter and the doors were nearly always kept open to allow access for the Lorries that continually brought in and took away large pieces of marine machinery.

By the time I had finished the job my hands were so wrinkled and soft from the continual immersion in the 'white water' cutting oil they resembled two lumps of shredded wheat. No doubt the next time the Donkey man sailed by the jetty he would corner someone and tell them that he knew the bloke who built it!

The *Rakaia* entered the locks to the Royal Group of Docks at around nine o'clock in the morning. As I was on watch I didn't see anything of our trip up to our berth which was right up at the top end of the Royal Albert Dock. We finally tied up outside the cold store and finished with engines was rung down at ten fifteen.

By the time I came up on deck the cargo hatches had been opened and the remainder of our cargo was in the process of being discharged. I was now on home ground, it was mid summer, the sun had come out and I had just completed

Back home in London's Royal Albert Dock, the ships on the right of the picture are berthed alongside what is now London's City Airport. (Photo John Layte).

a voyage right around the world plus the odd diversion on the way. In all we had travelled over thirty thousand miles, at times we had to work our guts out for every mile, these episodes would soon be forgotten, but the good times would be remembered for the rest of our lives.

Although we were back in the U.K. the voyage would not be officially finished until we paid off at North Shields in two days time. During the morning Mr. Strachan the Companies Marine Superintendent came aboard to see the Chief and collect a set of indicator cards that had taken Big Ron the 3rd Engineer, hours to prepare, these were for office use and had to be spotless with none of the usual splats of oil covering their entire surface. He also spoke to us to see how we were and to ask if anyone wanted to stay aboard for the next voyage. Everyone with the exception of the Chief Freezer and the 2nd Electrician requested to stay on for the next trip.

Later that afternoon I was able to nip home to see my family, it was only about three miles as the crow flies, but on the other side of the river, locally known as 'over the water'. As the voyage wasn't officially over I didn't have to pack my gear, I just took an overnight bag and the usual small souvenirs that seamen seem to collect on their wonderings around the world and except for the Pitcairn Island carvings, most of it was junk.

Due to strict union rules we were not allowed to do any repairs while in the London Docks, so my presence would not be missed. The Chief had given me the O.K. but as I went past his cabin he sang out,

"Dunni be late in the morning seventh, we are due to sail at midday."

When I arrived home I found my parents were expecting me. This was because my father who worked in the Royal Arsenal had seen the *Rakaia* come up the river and enter the locks from where he was working and assumed that I would be home later that day.

That evening I went out for a few drinks with some of my old friends and as usual on our nights out after the pubs closed we set off for London's West End. Well lubricated, we all piled into my friend Bernard's old van. It had been converted with seats and side windows which was a common practice in those days. As we came round a corner on a new one way system at New Cross, we were confronted by a car which was parked exactly on our predetermined course that should have navigated us round the bend in the shortest time.

Bernard solved the problem by side slamming it into the front of an adjacent house. As I was sitting on the near side I took the full impact. The side window shattered into my face, the van lost a front wing and several windows and acquired a new set of dents all down one side and a little later I acquired thirty five stitches. Fortunately no one else was hurt except perhaps the woman whose house now had a Vauxhall Wyvern parked up against her front window. She kindly came out with a towel to wrap around my head which by then was leaking blood in every direction making things look a lot worse than what they were.

As she handed me the towel the poor woman fainted, she must have seen that my eyebrow and part of my forehead had been almost sliced off and were hanging down over my cheek. She was still laying on the ground and being seen to by one of the ambulance staff as I was carted off to the Miller Hospital in Greenwich to get a new image.

The nurse told me that I was very lucky to have an African Doctor who had gained his expertise sewing people up during the Mau Mau troubles in Kenya. I remember her saying that the scars would eventually blend in as the years went by. Time has since proved her quite right as they have now joined up with several other scars that I have acquired over the years and all but disappeared among the wrinkles of advancing years.

One of my friends had stayed with me, so after being boxed up by the good Doctor we took a taxi home. Bernard on the other hand, had the van to deal with, not to mention the police.

Apparently when the police did arrive, the officer in charge asked him if it would still go, much to everyone's surprise when he pressed the starter button the engine started, the officer then said,

"I suggest that you put all the bits and pieces in the back and clear off."

Not believing his luck, Bernard, together with the rest of the lads beat a hasty retreat sans one front wing and windows. How times have changed!

When I woke up in the morning my pillow and sheets were covered in blood. My mother bought me in a cup of tea and promptly dropped it in shock. It wasn't until I looked in a mirror that I realised how bad I looked, I had two black eyes, one of which was swollen shut and besides the stitched up bit, my face had a multitude of small cuts. Compared to me, the Frankenstein monster would have won the world's most eligible bachelor competition hands down. The one thing we shared in common was that our bird pulling days were over, but I reflected on the fact that I still had my eyesight and realised how lucky I had been not to have lost an eye.

I was now in a predicament, as the *Rakaia* hadn't finished her voyage, there was a good chance that if I didn't get back to the ship it would sail without me and I would have a VNC (voyage not completed) stamped in my discharge book. Normally at the end of a voyage, providing you had behaved yourself it would be stamped VGC (very good conduct).

The first thing I said was "I've got to get back to the ship."

"You can't go anywhere looking like that, I thought we had finished with all this sort of thing, whatever happened to you this time," she asked.

"If I don't get back I'll have to go on leave, the last thing I want to do is stay at home looking like this for the next few weeks, I'll be O.K. once I get back on board." I replied.

I cleaned myself up as best I could and put a fresh bandage around my head but there was no getting away from the fact that I still resembled a Frankenstein creature, to add to this effect the whites of my eyes had now turned red.

After a cup of tea and a good breakfast a taxi arrived to take me back to the ship. When I got back on board I reported to the old Chief.

"Och aye you certainly made the most of your first night home, looks like a shark has had you by the head," he said as he smiled to himself.

Old Shipmates!
Top picture 1964. Bottom picture 2008 – Brian (The Bear) Bob (Eccles) & Bob (007).

I explained that I still wanted to stay on board for the dry docking and take my leave afterwards.

"It's fine by me laddie, but I'll have to see if the old man will O.K. it, go and rest up for a wee while and I'll see what I can do." He replied.

A short time later the Captain arrived accompanied by the Doc.

"I've seen worse, nothing like a good run ashore to keep up the spirit," he remarked.

The Doc gave me a quick check over and pronounced me fit for work as soon as I felt up to it. After they left I went along to the smoke room and gave the lads a good reason to think up plenty of ribald remarks which included, 'Didn't you know her old man wasn't at work' and 'Christ Davy, did you have to put yer whole head in'.

The Chief told me to rest up and he would arrange for the Fiver to stand my watch for me, this was the sort of ship board camaraderie that made the *Rakaia* a happy ship and often created friendships that would last for the rest of our lives. On long trips friendships were always made, but once the voyage was over, people changed ships and often lost touch, perhaps meeting up again on another ship, sometimes years later. These were often referred to as Board of Trade acquaintances.

Bernard, being concerned about my condition came round to my parent's house shortly after I had left for the ship.

"He's gone back to his boat," said my mother.

"But he looked like he was at deaths door when they took him away, is he coming back later," asked Bernie.

"No he's off to North Shields for a few weeks, he said he'd see you when he gets back," she replied.

Bernard later told me that he thought for a while, that perhaps he had imagined the previous night's escapades, but that was until in the light of day he saw the state that his van was in. It wasn't worth repairing so he drove it to the local scrap yard which was where he had bought it just a few weeks previously.

15
Into Dry Dock

We sailed for North Shields later that afternoon. On our arrival we were all paid off and the voyage was officially over, we had been away for five months and two days. After deductions I was paid off with one hundred and thirty-five pounds, seventeen shillings. I had also been sending home an allotment of £12 a month; this was to cover any expenses

Pay off document.

that I incurred while I was away. The regular ones being the hire purchase and garage rent on my vintage car.

The *Rakaia* then went straight into Smith's dry dock and once we had settled down we went on to shore power and our generators and boiler were shut down. A sudden uncanny silence pervaded the engine room. The heart seemed to go out of the huge main engine that had just taken us right around the world with only the occasional bit of trouble. Under these conditions the engine room took on an almost eerie kind of atmosphere.

Polished handrails that a few hours earlier were too hot to hold without a piece of rag became cold and damp from condensation and soon acquired a fine coating of rust. The slightest sound, such as a spanner being dropped caused echoes to reverberate all around the engine space in defiance of the overpowering silence. A dead engine room soon becomes a miserable place to work and the warm friendly atmosphere of a live engine room is soon replaced with the smell of stale gas oil and cold bilge water. Luckily this is a rare occurrence and only usually happens after a ship enters a dry dock for a survey or is laid up for a while.

Most of the lads left to start their leave and with the shore side engineers doing all the repairs, I had little to do. I made a rapid recovery, due no doubt to a good coating of diesel oil, which you never seem to get out of the pores of your skin.

After about a week I had to visit the local hospital to have my stitches removed. When the nurses at the reception desk asked me for my name and address I said I was in the Merchant Navy and my ship was the *Rakaia* berthed at Smiths Dock. They just nodded and smiled at each other with an all knowing type of grin.

I returned to the ship with the turban like bandage consigned to the medical waste bin and its place taken by a strip of plaster which now covered half my forehead. I had also been given instructions from a very nice looking Geordie nurse, which roughly translated meant that I had to keep it on for few days.

I was beginning to feel that I could now face the delights that North Shields had to offer, so that night I accompanied a couple of the lads just along the road to the Northumberland Hotel which was known by seafarers from around the world as 'The Jungle'. This was a tavern of ill repute if ever there was one, the best thing to do if you wanted to come out alive was to keep your head down and keep drinking. I don't think I understood a word that was spoken during all my visits to this infamous watering hole.

This became our local for the duration of our stay, but I have to say that I never experienced any trouble during our visits there; perhaps it was my looks that kept me safe. One night we decided to venture across the river on the ferry to South Shields, we had been told that there was a dance being held in a hall above a pub in the high street.

About ten o'clock I felt like getting back on board, perhaps I wasn't as fit as I thought, so I told the lads I would see them back on the ship. I had only walked a few yards down the street when I was confronted by two local hard cases. They seemed to appear out of nowhere and jostled me into a shop doorway.

"Let's have yer wallet." Said one

"And yer watch." The other said, as he pushed me back against the door.

It so happens that I never carried a wallet or wore a wrist watch. A seaman visits many dubious places in the course of his travels and ever since ships sailed the seas he has been a target for the dockside riff-raff and parasites. It was on the advice of the old hands that I never carried a wallet. If there was a likelihood of getting 'rolled' or mugged as it is termed today, it was always best to keep your money rolled up in your fist, which was exactly where mine was at that moment.

Watches and rings were a potential source of danger in the confines of an engine room. Rings could get caught up and the wearer might lose a finger or in the case of a wrist watch, if the strap got caught up it could pull you off balance which might lead to serious consequences. Neither had any practical use in an engine room!

I told my intrepid muggers that I had neither.

"Empty your pockets or we'll give you a hiding," I remember one saying.

It was quite obvious they were going to give me a good going over, so playing for time I made a start of going through my pockets. The last thing I wanted was to have my head split open again. For a second or two their attention was drawn to my coat pocket, it was now or never!

I hit the biggest one full on the nose with a punch that would have made a boxing coach keen to develop my pugilistic skills. Without waiting to see the results I hot footed it down the road as fast as I could towards the ferry. There was a considerable area of waste land that had no lighting, probably an old bomb site, that I had to cross to get to the ferry berth. I made it across in full flight without tripping and arrived at the quay just as the ferry was about to leave. As we pulled away I looked back for the first time, but could see no sign of my assailants so perhaps they had given me up as a bad job.

The next morning at smoko we had the post mortem on the previous night's events. When I told the lads what had happened to me and they saw my hand which by then had become swollen and quite black and hurt like hell they all had a good laugh and one said,

"You should stick to the ship Davy, every time you go ashore without us you end up as a hospital case."

"He's only done it so he can go back and see that Geordie nurse he was on about the other day," another added.

I decided to keep my shore going ventures to the now familiar 'Jungle'.

The survey work on the *Rakaia* progressed well; the propeller had been refitted after the shore gang had exposed the tailshaft for the surveyor and the stern gland had been repacked. The main engine was all boxed up and after we made a thorough check for foreign bodies, such as spanners and pieces of rag the crankcase doors were all replaced. The boiler had been examined and then flashed up and brought up to 120lbs per square inch. The safety valve was then set and tested to 100lbs/sq inch. Many other jobs were finished including a thorough cleaning of the scavenge trunking and spaces. Also the fresh and seawater pumps had passed the survey together with the bilge, ballast and lubricating oil pumps. The other and potentially dangerous items given a survey were the two large air receivers and compressors.

After three weeks we were ready to sail to London where we were due to load a general cargo for Samoa, Fiji and New Zealand. All in all we had enjoyed our stay in North Shields, the Geordies were a good work force and once I got to know and understand what they were saying I found them to be a great bunch of lads.

The day before we sailed the engineers who had been on leave returned and the stand by ones left. We had the same team except for the 2nd lecky and the junior, both had left to join other vessels in the Company. Once we docked in London I was going to have three weeks leave and as my face had practically healed up and I was looking a bit more like a human being I was looking forward to it.

The dry dock had been flooded and we were back on our own power enabling us to warm the engine through for twenty four hours before sailing, so the engine room became more homely after being shut down for nearly three weeks. We ran all four generators to check them over and to make sure all the systems were working. Once the 2nd engineer was satisfied three were shut down and everything was prepared for sea.

When the time came to leave all hands were on duty down below, a second generator had been put back on line and the main engine primed. The telegraph rang to stand by and the 5th engineer was on the controls, the Chief liked all of us to be experienced in manoeuvring the main engine.

I remember the first time I had a go at starting the main engine; it was an experience that I shall never forget. When the main lever was moved forward to the start position the rush of compressed air into the cylinders gave an impression of immense power. It wasn't possible to see the piston yokes rise and fall from the control position so the only way to judge when the engine had turned over was by looking at the link rod that drove the fuel circulating pump which was just to the left of the operator.

Once she was kicked over and the indicator cocks had been shut she was ready to start. The art was to start it on the minimum amount of compressed air without having a false start and having to do it again.

It was necessary to conserve the air as much as possible; this was to help the compressors keep up with the demand. There were times when the ship was having difficulties docking that we came perilously close to running out of air. When this happened we would phone the bridge and tell them that they had three starts left, there was always a bit more but we liked to give them something to think about.

As the engine started there was a heavy thumping sound from above as it fired along its sixteen combustion chambers. Once she was fired up it was surprisingly quiet and responsive for such a large piece of machinery. It was possible to go from full ahead to full astern as fast as you could move the levers. This of course would only be done in an extreme emergency. Under normal circumstances we would ease the engine through the rev range before manoeuvring.

For readers not familiar with large marine diesel engines perhaps I should explain that the majority were of the direct reversing type. This meant they didn't have a gearbox and clutch so to go from ahead to astern the engine had to be stopped. The camshaft was then moved along to bring in line the astern cams and the engine could then be restarted in the opposite direction on compressed air.

As it was the first start up after the survey all the engineers were on duty, even the Chief made an appearance wearing his boiler suit and carpet slippers. The telegraph suddenly rang and the pointer ended up at dead slow astern. The fiver calmly moved the levers and the huge dormant piece of machinery kicked over blowing jets of sparks out of the indicator cocks. Once the thumbs up signal came down that they were shut and all was well, the fiver pushed the fuel lever over to start, watching the fuel pump link rod rise and fall he pushed the lever further over to Run and the engine picked up straight away. He bought the rev counter back to 35rpm and stood back waiting for the next movement. Compared to the more well known Doxford engines, this type of Burmiester & Wain engine was very easy to manoeuvre it also had the advantage of not having to keep a constant eye on the fuel pressure.

We were under way! We could only visualize what was going on up top.

The telegraph rang to stop, we were probably clear of the dry dock and being assisted by a pair of tugs. A few more movements came before we got Half Ahead, so we were now probably out in the river and heading towards the sea. We had one more stop to drop the Pilot and we were on our way and as everything seemed to be running well all those not on watch left the engine room.

We had a minor incident during the trip down to London, the stern gland was found to be running hot. We slackened off the nuts and played a fire hose on it so that by the time it had settled down the tunnel took on the appearance of a Turkish bath. We emerged through the water tight door into the warmth of the engine room soaked to the skin and frozen to the bone. On reflection, the stern gland required no more adjustment until we reached the coast of New Zealand.

We entered the Thames estuary and as we passed Southend I noticed several ships laying at anchor, but this time we were lucky as we continued straight on past them towards Woolwich. Just after picking up the Pilot, the Chief came to see me and said

"Get your shore gear together, you can jump ashore when we get in the locks, we'll see you in three weeks."

This was a bit of a bonus as it could take an hour or more to get to our berth right up at the far end of the Royal Albert Dock, so I would be home before the *Rakaia* lowered her gangway.

16
The South Seas

My leave passed all too quickly, so I soon found myself reporting back on board to the familiar and homely surroundings of the *Rakaia*. I noticed she was flying the Blue Peter at her masthead and she was down to her Plimsoll mark so it looked as though we would soon be under way. Later that morning we signed articles in the saloon and a notice was put at the top of the gangway to say that shore leave ends at 1600hrs.

We sailed that night for the island of Curacao in the Dutch West Indies to take on fuel. Just over a week after leaving the British Coast the weather started to improve, the sea lost its angry green colour and changed almost overnight to rich tropical blue. We then went into whites as we entered the area known as the Bermuda Triangle some said that perhaps we would disappear and become another statistic. The strange thing was that there was an element of truth in that actually happening. We were at the height of the cold war and what we didn't know was that America was on the brink of wiping out Cuba and everything in the area with a nuclear bomb.

The sea was flat calm and we were now accompanied by dolphins and schools of flying fish, there were long streams of rust coloured seaweed appearing all round the ship so this indicated that we were in the infamous Sargasso Sea. We managed to collect some by throwing a small grappling hook over the side, the weed was in long lengths and on close examination we could see small crabs and other forms of small marine life in its fronds.

The Atlantic crossing was routine as far as the machinery was concerned but what was significant was the introduction of Uckers. Our new Junior Engineer, Tony, had spent several years in the Royal Navy, latterly as a Stoker in submarines. He certainly looked the part, as he sported a large black full set that would have made Blackbeard the pirate quite envious. He also had a typical submariner's sense of humour.

Uckers was a complicated version of the board game 'Ludo', any further affinity with this well known game ended as soon as the first dice was thrown. The main object, apart from winning was to gain the skill of being able to come up with a new rule that took advantage of the situation that presented itself with each throw of the dice. Once this skill was mastered you could become a top league player.

I can't recall where the 'Ludo' set came from, I can only assume that Tony bought it with him when he joined the ship, but there was a possibility that he found it, where it had lain dormant since the *Rakaia's* passenger ship days, in the locker at the base of the bookcase that we termed 'The Library'. The game took on a serious following among the Engineers; it even took precedence over cockroach racing. At first the Mates took a dim view of the whole thing, but soon became hooked once they had been officially challenged on the notice board.

All this interest meant that we had to make a winners cup, this was done in the engineers workshop utilising a Guinness can as the basis for a very professional looking trophy. By the time we reached Curacao the competition for winning it took on the significance of winning the 'Ashes'.

Our old Doctor and drinking companion had sadly left the ship at the end of the last voyage and been replaced by a younger more straight laced person who thought that mixing with the Engineers was below his standards so we saw very little of him. Later as we crossed the Pacific, he even managed to evade the administrations of King Neptune and his acolytes, I think a good seeing to by them would have improved his general acceptance of shipboard life no end. I only had one encounter with him throughout the whole voyage. This was shortly after we left London; he followed me into the smoke room after lunchtime and insisted on looking to see what lay beneath the plaster on my forehead.

I explained that everything was ok, but as he examined me he suddenly ripped the plaster off and in doing so opened up a bit of the scar tissue. He came close to receiving a Glaswegian Kiss, but he stepped back too quickly and immediately apologised. After that I wouldn't let him anywhere near me, the worst thing about it was, the blood ran down on to my shirt and for some reason I was unable to remove the stain, so I had to buy a new one from the Slop Chest. This little incident may have been the reason why he kept out of the Engineers way.

By the time we reached Curacao my wound had completely healed, I think the bronzing sessions in the Caribbean sun had a lot to do with it.

After we completed our bunkering and the spillages had been cleaned up we sailed for Colon and the transit of the Panama Canal.

We left the Miraflores Locks and entered the Pacific in a torrential downpour, the visibility was practically zero and the atmosphere was highly oppressive. We would be taking a more northerly route across the Pacific than on our previous return trip; we were due to pass close to the Marquesas Islands, then through the Tuamotu Archipelago with its numerous reefs and atolls to Apia in Western Samoa.

We cleared the bay of Panama and all was running well, I was sunbathing down aft on top of the spud locker when I was suddenly enveloped in a cloud of familiar smelling smoke, it was no use staying there so I headed back to the accommodation. As I walked back along the deck, huge billows of smoke were coming from the engine room access that led out on to the after deck and suddenly the sound of the engine room emergency alarm went off, indicating a bigger than average scavenge fire. It required the attention of all the engineers and greasers and for a time it looked serious, the paint on the top piston yokes was on fire with flames shooting several feet above the engine.

We kept cutting in and out units as and when required, hoping to keep the engine running. By the time everything was back under control I discovered that we had been in the engine room for seven hours. All the paint had been burnt off the top half of the engine and the upper part of the engine room casing and skylights had a thick covering of soot. The boat deck also received its share of black sooty flakes together with a scattering of lumps of coke. This turned out to be the worst scavenge fire that I experienced on the *Rakaia*.

141

Such was the intensity of the fire that if we had had a Doxford main engine it would have to have been stopped because the rubber hoses would have soon burnt through. Luckily the only damage incurred was to three of the cylinder lubricators where some of the sight glasses broke, something that was easily remedied.

It took several days to clean up the mess and get things back to a presentable condition. We resumed normal watches and started to look forward to our short stop over at Apia. The mate mentioned at breakfast that if there was any indication of bad weather we would have to by pass Samoa and continue on to Lautoka on the island of Viti Levu in the Fiji islands.

The problem was with the anchorage off Apia; it was very restricted and surrounded by coral reefs which made it a dangerous place to be if a strong wind and swell came up. Many ships have left there bones on the reefs. The weather held fine for us and we entered the lagoon and manoeuvred into a good position about half a mile from the shore. I was looking forward to a run ashore as this would be my first visit to a South Sea Island. It was arranged that shore leave would be split into two parties; this was to make sure that there was always enough officers on board to take the ship to sea if there was a sudden deterioration in the weather.

I was in the first group to go; we were picked up by a launch and taken to a small wooden jetty. I couldn't help noticing that our boatman was a rather fearsome looking character; he only wore a piece of coloured material in the manner of a skirt and a pair of sandals. He was heavily tattooed, the upper part of his legs were tattooed or dyed a pale shade of blue.

We soon discovered that there were no pubs on the island, apparently alcohol was forbidden to the natives. There was a Colonial Club just along the road, but we were told that it was for members only. I was more interested in having a look round and perhaps going for a swim. Just across the bay on a hill side, the tomb of Robert Louis Stevenson or 'Tusitala' (the teller of tales) as he was known as on the island, was pointed out to me.

A narrow road ran parallel with the beach and on the landward side the thick green jungle like foliage came right up to its edge. Dotted about among the palm trees were thatched dwellings with side curtains made from what looked like palm fronds. The majority of these were rolled up leaving the living area completely open and as we strolled about the inhabitants gave us friendly waves.

I noticed that several women were suffering from Elephantitis and had severe disfigurement of the legs. Many of them were also tattooed, but in a different pattern to the men. Out from the shore were long rickety looking, wooden jetties with a small hut on the seaward end, these I was told were the toilet facilities.

In view of this and the fact that the foreshore was more mud than sand we decided to give up the idea of going for a swim. It was also populated by a colony of ferocious looking crabs through which we would have had to have run the gauntlet to reach the water. There were two rivers in the vicinity so this probably accounted for the muddy beach.

We ventured along the coast and came back along an inland track surrounded by thick vegetation and occasionally catching a glimpse of the odd dwelling hidden among the trees. It was soon time to get back to the jetty and get the launch back to the *Rakaia*. All in all it was a bit of a disappointing visit; given more time to explore I'm sure we would have found one of those exotic beaches depicted in the Hollywood movies and perhaps a bevy of beautiful grass skirted Hula Hula girls to go with it.

I had a rare view of the *Rakaia* as we headed back in the launch, she was right down on her marks and had a purposeful look about her that gave the impression of seaworthiness. She had been visiting these waters for the last fifteen years and as she lay at anchor with the usual small column of black smoke from her generators rising out of her soot blackened funnel she appeared completely at home.

As the launch neared the ship I noticed how travel worn she looked. A group of Cadets were sitting on a plank that was suspended by ropes, they were painting out some of the rust streaks on the side of her hull and I could see Frank the Bosun leaning over the gunnels giving them a hard time with a mouthful of his colourful language.

Most of the consignment of cargo had been unloaded into lighters by the time we got back, but as we had to wait for the tide, the rest of the crew that wanted to go ashore had their chance. It was noticeable that none of the old hands bothered to take up the offer. In the mean time we had our dinner then went on watch and made ready for stand by.

A phenomenon that took some getting used to, was the loss of a day when outward bound to New Zealand, this occurred when we crossed the International Date Line just before reaching the Fiji Islands. The clocks had also been put back for the last week or so by about twenty minutes every twenty four hours, this rather complicated watch keeping times so that the Second Engineer had to organize a meal relief rotor to suit the changing times.

On our return voyage we would gain a day, so that if we crossed the date line, for example on a Thursday, the next day would also be a Thursday. After that we would have twenty or perhaps thirty minutes added every day until the time was adjusted to Eastern Time. This tended to give an impression of time travel while moving at a very slow speed, i.e. 14.5 knots. It's interesting to note that we never seemed to cross the Date Line in an Easterly direction on a Sunday. This would have given us an extra Sunday at sea and a few more Shillings in our pockets at the end of the trip.

We weighed anchor in the early evening while it was still light and headed for the Fiji Islands. Our next stop was at Lautoka where our arrival was greeted with the sound of South Sea Island music being played from loud speakers on the quayside. Once we were tied up we were boarded by huge Fijian wharfies who soon had the hatch boards off number five hold.

They were a happy looking crowd and they soon had some of the cargo swinging out on to the quay. However I couldn't help noticing that they frequently took a swig of something from a large jug that they kept within reach. The Donkey man duly informed me that it contained Kava Juice and was the staple drink of the Fijians.

They unloaded some general cargo and put aboard several tons of bagged sugar. Sugar was the main export from Lautoka and huge areas were given over to growing it. We only stayed for about five hours so we never had a chance to get ashore.

Our next port was Suva which was on the opposite side of the island. We stayed there for several days so we were able to have some good runs ashore and see something of the town. It was the time of the Hibiscus Festival so everyone seemed to be in a party mood; however after future visits I found that this was the normal happy outlook of the native Fijian people.

Our evenings were spent in the Golden Dragon, a seedy night club on the outskirts of town. One night we were invited to a Kava party which was held in a corrugated tin hut somewhere in the interior of the island. On entering it took a while to adjust to the dim lighting that was given off by two or three paraffin lamps.

There were three of us and we were indicated to join a group of Fijians who were sitting cross legged around a large bowl containing a liquid that made the washing up water in our galley look like spring water. A huge Fijian with an enormous head of tightly curled black hair leant forward and dipped both hands into the bowl and lifted out a tangled mass of fibres. He then proceeded to ring the mass out between his hands before replacing it and giving it a good stir. While he was doing this he looked in our direction and then gave a couple of slow nods of his head, accompanied by a knowing grin.

I was beginning to think that perhaps we should not have accepted their invitation, but it was apparent that we had no alternative, but to stay and see it through. What was noticeable and disappointing was that there were no females present, this rather dampened our enthusiasm, especially as our fearsome looking hosts not too distant relatives had been cannibals.

As the night wore on it became evident that my initial apprehension was quite unfounded, our new found friends were brilliant company. The downside of it all was that we had to drink in one go, the contents of a cup made from half a coconut shell that the Chief had filled from the large bowl.

This was the infamous Kava Juice that I had seen the natives swigging away at on the ship. After the first cup had been downed I found it tasted slightly better than it looked, it left a slight peppery flavour in my mouth. I was told by the Fijian sitting next to me that it was in fact made from the roots of a pepper plant and that in the old days it was chewed by the virgins of the tribe and then spat into the bowl. I began to wonder how old the 'old' days were!

The Kava session continued until the bowl was empty, I had heard that it was pretty potent stuff, but it seemed to have no effect on me or my companions. How wrong can you be! As I went to get up I found I couldn't move my legs, they were completely paralysed. The three of us were all the same; the Fijians seemed to be unaffected and found our condition highly amusing. They helped us outside where I was able to take some deep breaths; I remember that the air was very warm and heavily scented with vegetation.

Eventually a taxi arrived for us, driven by one of the Kava drinkers, in hindsight I thought that it was a bit strange, as all the taxi drivers, in fact anyone who did a normal days work were always Asian. Fijians only worked as and when absolutely necessary and seemed to enjoy life to the full.

Our driver took us back to the ship free of charge and even helped us up the gangway. The next morning I was none the worst for wear, no hangover, just a slight tingly feeling in my legs. However I decided to keep off the Kava in future. We left Suva the next day for Auckland in the North Island of New Zealand.

17
Sly Grogin

While serving my apprenticeship I had the misfortune to break a front tooth. After visiting the dentist he suggested that he could cut it off flush and fit a crown and that once fitted it would never drop out, so I took his advice and had it done.

The day after putting to sea following the Kava session, I was having a couple of Guinness with the Fourth just before lunch when the crown fell out! The Chief happened to be passing and found it all very amusing saying.

"Och ye should ne drink the Kava, its well noon it makes ye teeth droop oot."

"Not to worry," said the Fourth, "Araldite will soon fix it."

He was quite right! After lunch and a lot of leg pulling, I mixed up a small amount and applied it to the crown which had a small metal pin protruding from it and pressed it back in. They say, there's nothing as permanent as a temporary repair. My bit of D.I.Y. dentistry lasted until I visited my dentist some months later, where he proclaimed it to be an excellent job.

We arrived in Auckland during the morning so once again I was on watch and missed the panoramic view of the city as the ship came into dock. As usual after everything had been sorted we all had the rest of the day off, all that is except the Third, who as always did the first day as duty engineer.

After lunch we were off like long dogs and soon found ourselves in 'Ma Gleason's', a famous watering hole patronised in the main by British Merchant seamen. In those days in New Zealand the pubs all shut at 6 pm, so the locals who generally knocked off work at five o'clock, swarmed into the pubs and consumed vast quantities of beer in the hour before closing. This was known as 'The five o'clock swill'.

However it was possible once you learnt the ropes, to drink all through the evening. Once the publican got to recognise you, this was where my newly acquired scar over my eye started to show some positive advantages, a nod and a wink got you in for an illegal drinking session which was generally known as 'Sly Grogin'.

In places like Auckland and Wellington I was advised to check out the emergency exits, as now and again the police would make a raid. This was a highly entertaining experience, as the bar would always be full, just like any pub in London's East End on a Friday night would be. When the raid started there would be a sudden rush to escape the clutches of the law, some people went out through windows, others hid themselves in various parts of the building and some went out of the back door straight into the arms of the police. The ones hiding were usually rounded up after the initial panic had died down. I was advised by the old hands to wait until the first rush had subsided and then take the back door, as the police would then be busy rounding up the first wave of escapees.

In the smaller towns such as Napier and Nelson a different more civilised way of life prevailed. The police would pop in and let the publican know that they were going to raid during the evening. The licensing hours meant that there was a very good social home life, this was made evident by the invitations that we continually received to go to parties at the homes of many gorgeous New Zealand girls.

The locals always left the pubs at six o'clock with what was known as a carry out. This consisted of one or more glass flagons, each containing about a gallon of beer. These were usually tied together with a length of rope and carried over the shoulders. When we were invited to parties ashore we always took one or two cases of beer and perhaps a bottle of spirit with us, consequently we were never short of invitations.

On returning to the ship after my first night ashore, I was introduced to the 'White Lady'. This was the name of the pie stall that was situated by the entrance to the dock gates, practically at the bottom of Queen Street.

The proprietor earned a good living serving up delicious meat pies together with tea and coffee to seafarers returning from their forays ashore.

This establishment was a clone of a pie stall that I used to patronise during my apprenticeship days in the London Docks. This was known as 'Harriet Lanes' and was named after a woman who fell into the machinery in one of the Chicago meat works. There was also a similar one next to 'Monty's' in the Pyrmont area of Sydney. These were about the only shore side eating establishments that we patronised and when we did, it was in the early hours of the morning on our way back to the ship.

There was no need to eat out, as the food during my time on the *Rakaia* was always very good.

However not all ships were good feeders, some years later I was on a ship where the cook was an alcoholic, he was so bad, he was what we termed a 'Dipso'. One Xmas he imagined that the turkeys were crocodiles and that they were going to attack him, so he set about them with a meat cleaver and then threw them all overboard. We ended up with some sort of salad and burnt chips for our dinner.

The morning after my first run ashore in New Zealand, we turned to and started stripping out the pistons for re- ringing that had shown up on the indicator cards as being down on power. At 10:30 we had our 'Smoko' (tea break) in the small mess room at the top of the engine room. This was where the post mortem on the previous night's events took place. The only one who didn't have a tale to tell was Big Ron the 3rd Engineer this was because as always he had done the first night in port as the duty engineer.

I had been lucky as the 2nd Engineer had instructed the Junior to do the first week on nights, so this enabled me to make the most of my first week in New Zealand. As usual, when on night work the 2nd gave us a list of jobs that had to be done. They were usually small jobs such as repacking glands on the various pumps, or changing fuel and air start valves on the generators.

I always seemed to get the job of overhauling the stuffing boxes on the main engine piston rods. These were heavy items that had to be lowered with a certain amount of care. The packing rings were made of precision ground segments and were etched with a number that had to be matched with the corresponding number on the next segment. Their edges were razor sharp which resulted in numerous cuts to the tips of the fingers. This was a job that had to be done right; otherwise the gland would blow by as soon as the engine was started and I wouldn't be the most popular bloke in the alleyway.

We were due to stay in Auckland for ten days, during which time most of our general cargo from the U.K. was discharged, the remainder was to be unloaded in Napier and Lyttleton. A certain amount of frozen lamb was also loaded in Auckland, destined for the U.K. market.

As we were off watches and on day work it gave the opportunity to work alongside the engineers, electricians and freezers whom we didn't normally work with when at sea. Consequently the engine room became a place of intensive activity and no place for the unwary.

Maintenance of the main engine continued together with overhaul of the four generators. The work continued right up until sailing time, with the last of the main engine pistons being boxed up two hours before stand by was rung down.

In spite of all the work, we managed to have evenings, Saturday afternoons and Sundays off. A notice appeared on the board saying that a coach trip had been arranged by another ship for a run up to Rotorua and any one wanting to go should see the Chief Steward. I and quite a few of the lads from the *Rakaia* took advantage of this opportunity to see something of the wonders of New Zealand and we were certainly not disappointed.

While on the New Zealand coast, Saturday afternoon was the best time for going ashore. In Auckland the favourite place for us was the Great Northern Hotel which was just up Queens Street on the right. They always had a live band with dancing taking place until 6 pm and what made it particularly attractive to the likes of us was that it was also frequented by plenty of New Zealand's gorgeous looking girls.

Many a romance blossomed from here and when it came time to sail most of us were seriously in love. The old Chief was in his element at these times, especially with the first trippers, saying that there was a Shaw Savill boat docking soon after we were due to sail and that its crew would be taking good care of our new found loves.

An atmosphere of melancholy pervaded the accommodation for the next twenty four hours or so, but after a few hours ashore in our next port of call which was Napier everyone was in love all over again. Our next port was Lyttleton where the remainder of our general cargo from home was discharged.

While we were in Lyttleton another opportunity arose for another coach trip, this time down to Queenstown. This turned out to be superb day out, we crossed the Canterbury Plain and followed the snow capped Southern Alps all the way to Queenstown and the last part was along the shore of Lake Wakatipu. Our visit coincided with the town's centenary and to celebrate it, all the men had grown beards. As Tony the Junior was with us and still wearing his magnificent black 'fullset' and the fact that we were the first outsiders to visit them for a while we were welcomed with open arms by the local inhabitants. Apparently the road alongside the lake had been blocked by an avalanche or landslide.

Left to right, 2nd Sparks, The writer and Norman the 5th Engineer in pioneering tourist mode in Queenstown, South Island, New Zealand. Lake Wakatipu and the Southern Alps in the background.

For us engineers the day turned out to be a bit of a bus mans holiday. Alongside the quay was moored the service vessel T.S. *Earnslaw*, she toured the lake delivering the mail and other supplies to the homesteads on its shore. She was powered by two triple expansion steam engines of around 500 hp each and quite amazingly she was still coal fired. The last thing we expected to be doing that day when we left for Queenstown that morning was to be grovelling around an engine room miles from the sea. The old girl was something like fifty years old and looking a bit worse for wear, but I understand from my friends in New Zealand that she has since been restored and joined the modern tourist brigade.

After Lyttleton we continued around the coast to New Plymouth slowly filling our holds and then to Bluff right down at the bottom of the South Island to load our final consignment before heading home.

Quite often in the ports of Australia and New Zealand we would be visited by the local 'Sky Pilot' from the seaman's mission, they were always made welcome, no matter what beliefs we held and they usually joined us in a beer or two and often tipped us off as to the best places to go to meet the local talent. I always found the Seaman's missions such as 'The Flying Angel' very welcoming establishments and a complimentary part of the Merchant navy. They never thrust religion down your throat and made all nationalities welcome. One of the things I remember most about them was that they always knew how to make a good cup of tea.

Oh yes we did like our tea, but it wasn't easy to get a good cup when ashore in foreign places, so if the port had a 'Flying Angel' we often popped in for a brew and a game of billiards. Another welcome pair of visitors was a man and his wife who went round the ships with a mobile movie show. They carried all their gear around in an old Rolls Royce and as I had a similar car back home we soon became acquainted. After the film show they always joined me and a couple of the lads for drink and while we rambled on about old Rolls Royce's, the rest of the lads kept themselves occupied drooling over his wife.

18
Trouble Pending

Our trip across the Pacific was broken by a short stopover at Pitcairn Island. Having gained some experience from my previous visit, I had accumulated a good stock of ships soap, plus I had bought along some boxes of brass wood screws that I purchased before we left the U.K. These were highly prized items at Pitcairn so they gave me some good bargaining power.

We left Pitcairn with a good stock of fresh fruit and a variety of the islands insect life and reached Panama with very little mechanical trouble. The transit of the canal was straight forward and I even managed to have six hours off duty, so I was able to take in the ever changing scenery of the Isthmus.

After a stop for fuel at Curacao we headed into the Atlantic where we encountered some heavy weather which necessitated the use of the storm boards to the side of our bunks. With these in place it was almost impossible to be thrown out while asleep.

This trip our home port was Liverpool where we arrived after fourteen days of continual bad weather. We had been away for a total of five months and ten days. After we docked the Chief called me into his office to say that if I wanted to return for the next voyage I would be promoted to Sixth Engineer. This rather surprised me because I had not given any thought to promotion. Unfortunately my wages would remain the same, but it was one step nearer to a fourths job which was classed as a senior position and did pay a little bit more. In hindsight filthy lucre was not the driving force about going to sea, I am sure that most of lads that I sailed with all those years ago feel the same and would do it all over again if they were given the chance. It is interesting to note that I had earned a total of £409.10s.9d but after deductions I paid off with £162.9s.3d.

The main difference about the new position was that I would be with the Third Engineer on the 12 to 4 or graveyard watch as it was known, the other advantage was that I would not be on night work when abroad. After accepting the Chiefs offer he said,

"As soon as you have paid off you had best be away the noo, but be back in three weeks mind."

That afternoon I boarded the train at Lime Street and arrived at London's Euston station later that evening and arrived home just after the pubs had shut. To keep myself occupied during my leave I worked for an agency that provided drivers for various firms whose regular drivers had phoned in sick. I used to phone the agency at 9 am to see if I was required, if so I would go off in the Rolls to pick up the vehicle in question and deliver what ever the goods were that had already been loaded. This was before the days of the H.G.V. licence; consequently I got to drive all types of commercial vehicles from light vans to heavy goods. I used to get paid £3 a day, which was quite good money in those days.

All too soon it was time to report back on board. I arrived just in time for dinner and found some new faces in the smoke room. We had a new Chief Freezer, Eddie Moody, he came from Dartford in Kent and had served his time with J&E Hall & Co. the very womb of fridge machinery manufacturers. Eddie was no stranger to the *Rakaia* as he had done several trips aboard her in the past and he told me he was happy to be back aboard. We also had a new Junior, Tony the ex submariner came back as Seventh and Norman was Fifth. We also had a new Third and to complicate things a bit our Second had his wife along with him for the voyage. This meant that we would have to behave ourselves in the accommodation and to a certain extent out on deck.

Brian the Third came from Liverpool and we got on well from the start and as I was going to be on watch with him this was an advantage to us both. He told me that he had never sailed on a ship with Burmiester & Wain main engine and had heard all about the *Rakaia's* troublesome reputation.

The *Rakaia* was the only ship in The New Zealand Shipping Company to be powered by one of these complicated looking monsters, so this and the reputation for generator troubles put the ship at the bottom of the preferred ships to sail on table as far as the Company engineers were concerned.

Although she was a hard ship to work, I thought that that she didn't deserve the reputation that she had. All the Companies ships had their share of mechanical problems, some more than others, but I have to say that during my time aboard her and travelling over 150,000 miles, the *Rakaia* only 'failed to proceed' while on passage on three occasions and these were only for short periods.

Although she might have been considered to be a hard ship to work on, she was always a happy ship. I am sure this was due to the old Chief Engineer John (Jock) Cowper, he always stood up for his engineers and made sure they were treated with the respect they deserved. However he never suffered fools lightly and expected all his staff to do their jobs in a professional manner. He also made sure that when in foreign ports we didn't have to work unnecessarily, such as Saturday afternoons and Sundays.

There were times when we had to turn to owing to engine problems and the need to maintain sailing schedules. On these occasions everyone accepted the fact that we had to work around the clock to complete the work in hand, even if it meant forgoing our shore side assignations with the fairer sex. While on the subject of sleep! We sometimes took turns to catch fifteen or so minutes of it on the engine room plates. This enabled us to keep going until some sort of normality was regained and leaving only the 'just jobs' to be sorted.

We sailed the day after I returned from leave, so unfortunately I didn't get time to see much of Liverpool. Apparently quite a bit of work had been carried out by the shore gang on both the main engine and two of the generators so it looked like we might have a have a trouble free trip, which I soon discovered was back out to New Zealand via the Panama Canal.

When Stand By was rung down the 2nd took the controls and gave the 3rd a quick course of instruction. By the time we dropped the pilot he soon had the hang of it and said they were much simpler to operate than on his previous ship the *Haparangi* which had Doxford engines. He gradually brought the main engine up to its most economical speed of 102 rpm which gave the ship a speed of around 14.5 knots.

We all stayed below for the next two hours and when the 2nd was satisfied that everything was running O.K. we went onto normal watches of four on eight off. As I got used to my new watch, the twelve to four, I found I preferred it to my previous eight to twelve. The midnight to 4.00 am or graveyard watch was the best, with only the Third, myself and a Greaser down below and the Second Mate and his assistant up on the bridge. The rest of the crew would be in their cabins and most of them asleep. The greasers and engine room labourers sometimes stayed up late playing cards in their mess room, but generally the ship had the feeling of being deserted.

The worst part of going on watch at midnight was when half asleep you opened the engine room door. The heat and noise hit you like a blast from an explosion and instantly bought you back to reality. After descending the several flights of steps to the control platform and relieving the previous watch, a mug of boiling hot tea soon fortified us for the four hours ahead.

19
Earning our Keep

We were two days out into the Atlantic with the usual heavy sea running, when I was suddenly woken up at about 10.00 pm by an unbelievable noise and a vibration that almost threw me out of my bunk. For a second or two I thought I was in some sort of dream, but it soon turned into a real life nightmare.

The engine room emergency alarm started screaming out in the alleyway and immediately bought me back to my senses. I rushed out of my cabin instinctively grabbing my torch on the way and met the 2nd and the Chief who said,

"Take care and dunni roosh doon below, we dunni noo what we'll find."

He was in his pyjamas while the rest of us were in our Y-fronts, by this time we had been joined by the Freezers and the Lecky's. As I was the nearest I pulled open the engine room door and immediately the alleyway filled with acrid smelling smoke.

"She's ne a scavenge fire, gets some wet rags and wrap them roond yer heeds," shouted the Chief.

He had to shout because the noise from the engine room was deafening.

As we felt our way down below, I noticed large sparks arching about in the direction of the main switchboard.

The visibility improved once we reached the controls at the bottom of the engine room. The Chief bought the main engine to a stop while I shut in the coolers. The 7th, who was on watch appeared from around the front of the engine and said that the 4th was in trouble as he was trapped behind number four generator which was still screaming away.

The 3rd and I rushed round to the port side of the engine room and were confronted by the worst scene of mechanical devastation that I had ever set eyes on. The armature of number four generator was a huge ball of flame and was still revolving and sending a shower of sparks across to the ships side. The engine room plates were covered in oil and on fire in the area affected by the fire ball from the armature. What was left of the engine was also still revolving with an occasional piece dismembering itself to fly haphazardly into the air until crashing into its surroundings.

Trapped behind all this was the 4th, he was vomiting into the bilge, probably due to a combination of fear and the inhalation of the acrid smelling smoke that filled the engine room. Luckily there was a very large foam fire extinguisher on its own trolley that was sited on the Port side of the engine, Brian the 3rd Engineer ran out the hose while I opened up the valve. A large stream of foam poured out of the nozzle and we soon put out the fire so that we could get to the 4th and pull him out. As we were doing this, all the lights went out and the remains of the generator came to a stand still leaving the engine room strangely silent and in pitch darkness. Luckily instinct must have kicked in when I had rushed out of my cabin as I realised that I had my torch with me, I also noticed Brian had his as well.

I stayed with the 4th while Brian went off to see what was happening, he returned and said the Chief wanted me to stay where I was and make sure the fire didn't break out again and to keep an eye on the 4th. He then said that he was going to start up number three generator to see if we could get some power back on.

Having lost way the ship was rolling quite badly, so it was almost impossible to get about in the area where I was owing to the oil and foam that covered the plates, so the 4th and I had to keep hold of the guard rail around the armature end of the generator. There was no point in trying to go anywhere as our shoes were soaking in oil, making it dangerous to move and do anything more constructive than to stand by with the fire hose.

While Brian started up number three generator I pointed the beam of my torch in his direction to help him see what he was doing, once he had it going he gave me the thumbs up and disappeared round the front of the main engine to hopefully put it on the switchboard. It seemed ages but after about fifteen minutes the lights came back on.

The first priority was to get the ship underway, the Chief phoned the bridge and told them we would be about another fifteen minutes as the Lecky's had to do some temporary repairs to the switch board. In the meantime we made a quick assessment of the remains of number four generator.

After a mug of tea the 4th returned to normal and gave us a run down on what had happened. The 7th had reported to him that the generator seemed to be revving faster than normal, he immediately checked it out and could find nothing wrong but he told the 7th to keep an eye on it for a while and returned to overhauling some spare generator fuel valves. After about ten minutes the 7th came rushing round and said that the geny was in trouble.

By the time they got round to the other side of the engine room it was running at twice its normal speed and seemed to be increasing, the 4th tried to slow it down by moving the fuel lever but it had no effect, the rev's just kept on increasing. He shouted to the 7th to sound the alarm while he tried to stop the engine. He disconnected the fuel supply but it still continued to speed up, by then it was revving in the region of 2000rpm and drawing up its lub oil as fuel and well on the way to self destruct. Unfortunately it was not possible to shut off the air supply so it kept on speeding up.

It then started to shed pieces of the valve gear so that in theory it was impossible for it to run. But by then the main switch board was suffering from overloading, resulting in a change of polarity to the armature so that it became an electric motor. It was this that kept the remains of the diesel engine spinning.

I found out afterwards that the Chief Lecky had to use a sledge hammer on the breaker switch as it had partly welded itself together, once he isolated the runaway geny from the board it ground to a stop. While I was standing by with the fire extinguisher I experienced what can only be described as a severe buttock clencher as I suddenly recalled a similar incident that occurred while I was serving my apprenticeship.

I only saw the results of the incident the next day as I walked along the side of the dry dock in London's King George V dock.

The ship in the dry dock had a large hole in its side some way below the water line. Not an unusual sight in itself, with a ship being worked on in a dry dock, but what puzzled me was that the plate work around the hole was bent outwards.

It transpired that the ship was preparing to go on to shore power when one of the ships generators had run away and while the 5th engineer was trying to stop it the flywheel exploded. A piece sliced off a large part of his buttock which hospitalised him for some time afterwards and a much bigger piece weighing over a hundred weight went right through the side of the ship and landed on the skin floor of the dry dock. The consequences of a similar disaster happening to us way out in the Atlantic didn't bear thinking about, no wonder the 4th had been spewing up!

Our two Lecky's checked out the main switch board and carried out some temporary repairs then gave the thumbs up sign to the 2nd Engineer who was standing by at the controls. The Chief phoned the bridge that we were ready to start the main engine and almost immediately Stand By was rung down on the telegraph followed by Full Ahead. During this time Brian had started a second generator and the Chief lecky put it on the board and distributed the load between the two generators to 600amps each.

Once we got underway conditions down below greatly improved. The engine room labourers were turned to on overtime to clear up the oil and mess around the broken generator, while us engineers who weren't entitled to overtime went on to six hour watches and were assisted by the Lecky's. Once the Chief was satisfied that everything was running well, we started to assess what could be salvaged from the mangled remains of the generator. While doing this it was generally agreed that it would keep us in Panama and give us plenty of time to sample the delights that the flesh pots of Colon had to offer.

What we hadn't taken into account of was the dogged determination and professionalism of the old Chief. The next morning he called me into his cabin as I passed by on my way to breakfast. It was to check that I wasn't suffering from all the fumes and smoke from the previous nights work and he added,

"Aye all the rest of the lads seem O.K., there'll be plenty to keep you all occupied over the next few days and with a we bit of luck we should have the geny up and running by the time we reach Panama."

After breakfast I went back down below to do my six hour stint and found that the disaster area had been cleaned up and the Second, Third and Chief lecky were sorting out what could be reclaimed and what replacement parts were required. Because the generators were a constant source of trouble the ship carried a large amount of spares, both new and used, so the viability of the rebuild depended on the unknown condition of the spare armature. This was secured by large brackets at the very top of the engine room just below the skylight and completely out of reach. The other doubtful thing was the condition of the crankshaft for which we had no spare.

Although the spare armature was a huge piece of machinery, its presence had gone unnoticed and had the appearance of being there since the ship was launched. It was completely wrapped and sealed in canvas and a cursory inspection from the engine room skylight revealed that its top surface was covered in a thick cocktail of asbestos and carbonised dust.

This had to be cleaned up by the engine room labourers before attempting to get the armature down to the top plates so that the Lecky's could give it the once over. Given the circumstances the labourers did a good job, but with the ship rolling and pitching it was inevitable that a lot of the dust drifted down through the engine room.

The big problem was that it was mounted above the travelling crane, although we called it a crane it was in fact a hoist and all engine rooms had a similar arrangement. The electric motor and winding gear were built as a unit and could travel along the length and width of the engine room supported on guide rails. It was controlled by an umbilical cable with a hand held control box, but in this instance it would be of no use to us until we succeeded in lowering the spare armature down to the top plates.

The use of chain blocks was ruled out because of height restrictions above its mounting point, so the chief came up with the idea of using one of the cargo winches. After liaising with the First Mate and the Bosun it was considered the best option. The deck cadets were put to work to rig snatch blocks at appropriate positions so that the wire from one of the winches that served number four hold could be run across the deck then up over the accommodation to the boat deck and through the skylight and down to the armature.

The winch at number four hold was chosen because the winch operator had a direct line of vision with the engine room door that led out on to the after deck, a rather unusual feature, but a positive bonus in our present predicament. The third Engineer was lowered down through the skylight in a bosun's chair so that he could make a sling out of the wire strops that were passed down to him. When all was ready, the weight was taken on the winch wire, via signals relayed to Frank the Bosun who was on the winch.

The 3rd then removed the bolts to the brackets that secured the armature, leaving it sitting in its mounting plinth; the large hoop shaped brackets were then lowered by ropes from the skylight down on to the top plates.

The next part of the operation had to be carried out to coincide with the rolling of the ship and depended on a bit of luck and good judgement. The movement of the ship had been reduced as much as possible by the bridge putting the ship on the best course and requesting a reduction in engine revs.

At what seemed the best time the Second Engineer signalled through to the bosun, the response was surprisingly quick, the armature was lifted clear and as the ship rolled to port the 2nd signalled for it to be lowered.

We had belaying ropes from the armature that went to round turns on suitable pieces of pipe work so we could guide its descent down to the top plates. The operation was a complete success and the armature was soon settled on a cradle that had been quickly made by the Chippy. It had been a superb co-operative effort by all the ships company, including the catering department who kept us supplied with tea and tabnabs.

The removal of the canvas wrapping was akin to the exposing of an Egyptian Mummy, we just didn't know what we were going to find!

Eventually all the protective material was removed, the commutator even had wooden slats wired in position around its circumference which indicated that it had never been used, it also had the added bonus of being fitted with new bearings.

A thorough check was then made by the Lecky's and in the meantime we were kept occupied settling in the main engine once the ship had been bought back on course. Our chances of a stop over at Panama were drastically reduced when the armature was given a good bill of health. The next big problem was to get the armature down to the bottom of the engine room and round the forward end of the main engine to the port side.

The only way was to remove the main set of steps and associated gratings and various lengths of pipe work, some of which had to be temporally by passed until we had the armature safely down on to the engine room plates. Prior to lowering, it was wrapped in old blankets donated by the Chief Steward and finally in some heavy canvas that we managed to scrounge off the Bosun, with strict orders that it was to be returned in the same condition and not covered in black oil.

Our next problem was to manoeuvre with a series of chain blocks the very large but fragile piece of machinery around the forward part of the main engine and down the port side past number three generator to its new home. To save time it was decided to leave the old burnt out armature down below until it was convenient to bring it up. This together with all the goings on around the broken geny, resulted in a very congested situation.

Once the spare armature had been declared fit for service the crankshaft was inspected, the journals were in a bad way, but after hours of filing, stoning and finally lapping we managed to restore them to a serviceable condition. In spite of having a plentiful supply of spare bearing shells, it was found that we were one short to suit the reduced diameter of one of the journals.

The only way to overcome the problem was to melt out the white metal from one of the spare ones and add some of the white metal reclaimed from the crankcase and then recast it back onto the shells. The Chief asked us if anyone had ever done this sort of thing before and as nobody put their hand up I ventured that I had done a bit of white metaling during my apprenticeship.

"Och well done 'Sixer' I'll leave you te sort it oot the noo," said the Chief as he disappeared towards the Galley.

He returned a few minutes later with two large ladles and handed them to me saying, "You'll be needing these, take good care of them mind, as Cook wants them back when your doon."

We had no gas on board but we did have a very big old blowlamp tucked away in one of the lockers in the workshop. When I checked it over I found that it still contained a small quantity of paraffin, but not enough for the job in hand, so I went off in search of Lampy and managed to persuade him to fill it for me.

The black art of white metaling is not a skilful job; in fact at The London Graving Dock where I served my time it was done by 'Little Arthur' one of the labourers. Apart from nearly setting fire to the workshop when the blowlamp decided to change into a flame thrower the job went well. With the pair of shells still clamped together I set it up in the lathe and bored it out leaving a minimum amount of white metal to be blued in.

While I was busy on the bearing the rest of the lads had been occupied dressing up the crankshaft to a satisfactory finish. After fitting a new set of main bearings a reasonable set of crankshaft deflections were obtained prior to chain blocking the new armature into position and bolting it up to the flywheel.

The one advantage of having troublesome engines is that the ship acquires more than its normal stock of spare parts. The *Rakaia's* engine room was no exception; every useful space was used to store them. Many, such as piston assemblies had seen service before, but could still be used in an emergency. The same went for cylinder liners; these were sometimes removed because of cooling water leakage around the sealing rings. When in dire need they could be refitted with new rings and suitable application of 'Araldite'. There's an old saying that suited our situation quite well, 'If you can't fix it with Duct Tape, you are not using enough Duct Tape'.

Where we did have a problem was with the push rods, these were items that rarely needed replacing, therefore we only had two spare ones. This left us having to reclaim ten of the bent ones. At about a metre long and 35mm in diameter it is hard to imagine how they could bend. The fact that they did, gives some idea of the revs that the engine must have been doing.

The generators installed on the *Rakaia* were not fitted with rocker covers so all the valve gear was exposed. Once the engine revs reached self destruct level, the push rods were shot into the air with such velocity that most of them hit the deck head some eight feet above. The remainder went up the side of the main engine, two were found in 'Hells Alleyway' between the exhaust trunkings.

The pushrods were straightened to the best of our ability by clamping them in the chuck on the old lathe. Then using a length of heavy tubing they were slowly trued up by applying various degrees of pressure at the appropriate places until we achieved a satisfactory result. Finally, the day before reaching Curacao the generator was given its test run.

I was looking forward to our short stop at Caracas Bay, on my previous visits I had always been on bunkering duty so this was my first chance to get ashore there. Unfortunately there would not be time to visit the town of Willemstad which was about five miles to the West, so I made do with a swim in the wired off area known as the pool, followed by a walk up to Captain Morgan's Castle and absorbed a bit of nautical history. The thing I remember most about my first visit ashore was the abundance of lizards that darted about on the sun drenched rocks, they moved with such speed they were impossible catch.

Five hours after docking we were under way with full bunkers heading for Colon and the Panama Canal. Our hopes of some time ashore there were well and truly dashed, as by now the rebuilt generator was on load and running well and the fact that we had saved the Company an enormous bill at the hands of the American engineers at the Canal Zone, justified our salaries for many trips to come.

The results of the inquest on the cause of the disaster was carried out with the assistance of several cases of beer and concluded that a distance sleeve in the governor housing had been fitted by the shore fitters on the wrong side of the main spring. The generator had only been run on test in the calm waters of the dock but once it had been put on load in heavy seas the governor was unable to control the fuel delivery from the pumps. This set of the chain reaction that wrecked the machine.

The Canal transit was just as interesting as ever and we were soon underway for the Pacific crossing. This trip we were taking the direct route to New Zealand and expected to reach Auckland in three weeks providing we had no further troubles.

We settled in to our routine of four on and eight off and enjoyed the facilities of the swimming pool and the deck games and the morning bronzing sessions down aft on top of the spud locker. Once we reached the southerly latitudes we would suddenly be joined by a solitary Albatross that kept station almost within touching distance from our Starboard quarter for days on end. Did these mysterious and lonely creatures patrol a certain area of ocean waiting their turn to escort the first ship to pass just for company?

Now that I was on the twelve to four watch, I was able to indulge in the Saturday evening film show out on deck. The projector was operated by the two Radio Officers or 'Sparks' as they were more commonly known. The main part of the entertainment was not necessarily the film, but the times when it came on inverted or broke down. The whole show was accompanied by the continual pop and hiss of beer cans being opened and when the projector played up the Sparks were showered with empty cans and a chorus of whistles and boo's. After they sorted out the problem and the film came on again it would be greeted by load cheers and another shower of empty cans. Perhaps all this strange human activity was what made our Albatross escort stay with us.

We arrived at Auckland after an almost trouble free voyage, the only significant incident was when the engine gassed up during my afternoon watch. Burmiester & Wain double acting two strokes suffered with this phenomenon on the odd occasion. It was caused by the combustion pressure finding its way back through a fuel valve into the fuel system.

The first symptoms were a gradual slowing down of the engine, sometimes if the offending valve could be found quick enough, it could be bled of air and the situation alleviated. The safest way of dealing with the problem was to shut in the coolers before the engine stopped. The fuel system could then be primed by using the hand pump and the engine restarted. The downside of this was that all the engineers would have to turn to for the duration.

There was another sad incident which was unrelated to the normal day to day running of the ship. We were taking two dogs out to join their owners in New Zealand and they were housed in purpose built kennels up on the boat deck. They were taken for walks around the decks twice a day by the Captain and became firm favourites with all the crew.

One morning in mid Pacific he went to give them their exercise and found that one had died in the night. The ships doctor, wearing his vetinary hat was of the opinion that it had died from the tropical heat. I had my doubts about his diagnosis as I knew that they were well cared for and always had plenty of water. It was sewn in canvas together with a scrap main engine bottom end bolt and buried at sea later that afternoon.

20
On the Coast

We started our routine engine room maintenance the day after docking in Auckland. By the time we had visited all the loading ports around the coast all the machinery would have been bought up to the best condition possible in preparation for the long voyage home. As was usually the case, the last hour before sailing the clonking of 28 or 56lb hammers knocking up large nuts could be heard emanating from deep down in the engine room.

With most of our cargo discharged we left Auckland to begin our loading program which would take us to Wellington, Lyttelton, Timaru, Bluff, Napier and then up to Opua in the Bay of Islands to load several tons of powdered milk and a consignment of lamb from a railhead at the end of a small jetty. Except for a small general type of stores, that never seemed to be open, there was nothing else there. On a Saturday afternoon the Third and myself caught a very infrequent vintage bus into the nearest town of Kawakawa

The *Rakaia* loading frozen lamb at Opua, Bay of Islands, NZ.

which was about eight miles away to check out the local talent, the bus also doubled up as the mail van as the driver kept stopping to drop off letters to the remote homesteads on route.

The town itself was very small, consisting of the main street which had a railway line running down the middle and a few minor roads leading off. At one end of the street was a hotel and bar and at the other was a down at the heel pub with sawdust on the floor and what appeared to be spittoons in the corners, all very Wild West in atmosphere.

We gave the small bar of the hotel a try first; it was full of locals all intently listening to the radio, as we walked up to the bar and ordered a couple of beers they all turned towards us and waved their arms up and down and made shooshing noises indicating for us to be quiet. The bartender put his finger to his lips, gritted his teeth then silently passed two beers across the bar to us. It soon became evident that his regulars were listening to a horse racing commentary and silence would have to prevail until the end of the race. When the winner was announced there was lots of cheering and back slapping and beers were ordered all round and with so much noise it became impossible to hold a conversation.

About ten minutes later the radio commentator announced the start of the next race and the bar full of punters all looked towards the radio and silence descended once more. Brian gave me a nod and we quietly slipped out the door, leaving the equestrian aficionados to enjoy their Saturday afternoons entertainment.

"Not one bit of crumpet in the whole place." Brian grumbled as we headed for the pub at the other end of town.

As we walked alongside the rusty looking railway line a strange rumbling sound came up behind us. It turned out to be one of those lineman's trucks, on board were two unsavoury looking characters pumping away at the cranking handles. The truck was loaded with sacks of what looked like provisions and as it trundled away into the distance Brian said they were probably gold prospectors. I had a strange feeling that we had somehow stepped back into the past!

Our new found watering hole was packed, the majority of the clientele were Maoris and they were obviously intent on having a good time. We were soon welcomed into their midst and after the pub shut we were invited to join them at a party at in an old colonial style bungalow on the outskirts of town.

It turned out to be one of the best shore going forays of the trip, we arrived back at the ship in the early hours of the morning in the back of an old truck accompanied by half the tribe who in between swigs of beer were all singing their entire repertoire of Maori war songs. As the truck skidded to a halt amid a cloud of dust at the bottom of the gangway the Cadet on duty must have thought it was some sort of raid. By the time we reached the top of the gangway and waved our new found friend's good- bye he had returned with reinforcements in the shape of the Third Mate. He sized up the situation straight away and disappeared back into the accommodation mumbling 'Bloody Engineers'.

The Bay of Islands was made a popular spot for big game fishing by the writer Zane Grey and was one of the main areas of the Maori wars. One day we took our crash boat over to Russell, once the capital of New Zealand and the place where the British signed a peace treaty with the unbeaten Maoris. It was once known as The Hell Hole of the Pacific due to its violent reputation but we found it to be a very quiet and friendly town.

These sorts of trips helped to keep the engine in a good reliable condition, it was the 5th Engineer, or 'Fiver' as he was known, who officially in his free time looked after the engine in the crash boat. However when the *Rakaia* was on passage through the tropical calms, he would usually be given moral support and adequate liquid refreshment by a couple of the lads who were also off watch.

The crash boat was kept up on the boat deck along with the life boats, one of which also had an engine. It was not unusual to find that seawater had found its way into their engines after encountering heavy seas in the Atlantic and the Southern Ocean. Our little trips in them not only helped keep them in good order, it gave the Engineers the opportunity to familiarise themselves with the starting and management of them. This was obviously good practice as we never knew when they might have to be used in earnest.

As was often the case, we returned to ports that we had visited only two or three weeks previously, this enabled us to renew our romances with the local girls who thought we had deserted them forever. My two favourite places were Napier and Nelson, both populated with an abundance of beautiful girls all of whom seemed to have a built in preference for British Merchant Seaman. It was inevitable that by the time we put to sea we would all be deeply in love again. It also meant that by the time we reached Panama our mail would contain our usual quota of 'Dear Johns' to provide a bit of on board entertainment.

One of the ports that we returned to after a two week absence was Lyttelton, it was here that Brian the Third and I had a narrow escape from the brothers of a couple of girls that we had entertained on board on our previous visit. They stormed up the gangway, pushed past the Cadet on gangway duty and shouted that they had come to sort out the Third and the Sixth engineers.

As I was duty engineer I happened to be down in the engine room checking that all was well, when the phone rang.

The 2nd Freezer contemplating his future after receiving a Dear John.

"It's me Brian, there's a couple of gorillas up here looking for us; I'm going to lock myself in my cabin. If they come down the engine room don't let on you're the Sixer, if they ask where we are, say that everyone is ashore."

Sure enough, they did find their way down below and sure enough they were a couple of gorillas, they could easily have been part of the New Zealand 'All Blacks'. It was by sheer luck that I was wearing a brilliant white boiler suit, so as soon as Brian rang off, I hot footed it up into the fridge machinery room in the hope of keeping a low profile.

It wasn't long before the irate pair entered and found me chalking up some fictitious notes on the notice board.

"Hello there, I'm the Third Freezer, can I help you?" I asked.

It was at that point that it occurred to me that the girls might have mentioned to them that one of us had a scar over his eye, so I did my best to keep my left side obscured while I kept chalking on the board.

"We're looking for the Third and the Sixth engineers," one of them scowled.

"They've gone into town with the rest of the lads, you'll find them in one of the pubs, can I give them a message?" I asked without looking round.

"Just tell them we are going to cut off their balls for what they did to our sisters," said the other one, as thankfully, they turned and left.

I stayed below for the next hour, before venturing up to ask the duty Cadet if the coast was clear. Having ascertained that they had left the ship I went along to Brian's cabin and gave him the all clear.

"Thank Christ for that, for a while I thought they were going to break my door down," said Brian, adding, "I wonder why they were so upset."

We agreed that it would be prudent to stay on board for the rest of our time in Lyttelton.

While in Napier we were invited to several shore side parties where we all had a great time. One however stands out in my memory because of an amusing incident occasioned by

Tony our ex-submariner. It was held in an old bungalow on the outskirts of town. Around midnight we started to run low on drinks, so to keep the party in full swing a couple of the lads took a taxi back to the ship for fresh supplies. They returned about an hour later with enough to see us through until dawn.

Once re-lubricated, Tony decided to demonstrate one of his special dances in the bungalows large hallway. He had a towel wrapped around his head in the style of a turban and was stripped to the waist. With his large black beard and swarthy completion he resembled a bandit from the Khyber Pass. As the dance reached its climax following a leap in the air, he suddenly disappeared amid a cloud of dust through the floor.

It was at that moment our host's mother came in through the front door, the sight of Tony up to his waist in the floor was too much for her and she fainted. This heralded the end of the party, while our host and some of her friends attended to the mother we managed to extricate Tony and beat a hasty retreat via the back door. As we made our way down the hill towards the seafront I was surprised to see a real life chain gang, I found out afterwards that they were from the local prison and were doing some work to the sidewalks.

The next day we managed to scrounge a selection of dunnage and a wood saw off Chippy. Together with a large bunch of flowers from a nearby florist we returned and made our apologies and set to and repaired the hole in the floor. In true N.Z. style the mother took it in good heart and we left on the best of terms, so good in fact that the Third arranged to take her out to dinner that same evening!

The *Rakaia* was officially adopted by the town of Nelson and every time we visited, there would be a pipe band playing on the quay as we came alongside. The ship also held a dance on the boat deck where all the towns' dignitaries together with their daughters were invited. In the early 1960s the *Rakaia* was one of the biggest ships to visit the town, this was because of the limited depth of water over the sand bar which surrounded the port so we practically had the girls to ourselves.

Her visits were treated as part of the social calendar by the good people of Nelson and we were spoilt for choice with invitations to join them for shore side parties and other nocturnal activities. There was a mental hospital on the outskirts of town staffed by a great crowd of nurses that we immediately fell in love with. They seemed to have a certain kind of empathy towards us engineers and gave us their full and undivided attention. Another feature of town was the Nelson Marine Services Association, they organised sight seeing trips for members of ships crews who were free to take advantage of their hospitality.

During my time on the *Rakaia* the Chief, Jock Cowper constructed a very accurate scale model of the ship. It was not uncommon to be stopped by him as you passed his cabin and be given a working drawing, such as a winch drum, saying,

"Ah 'Sixer', do you think you could do me a favour and do me a wee bit of turning the noo."

What he in fact meant was that he expected me to spend several hours on the old lathe down in the engine room workshop machining up small bits and pieces for his model.

I was not the only one; I think he conscripted most of the crew at one time or other. He discovered that one of the engine room Greasers was in fact a skilled watchmaker, so he was given the task of making the anchor chains and various other small components. The finished model was a piece of first class model engineering and could easily have taken a gold medal at London's Model Engineering Exhibition.

The Captain suggested that as it was to the same scale as the Company's model that was kept in a glass case in the cadets anti room, it should be exchanged and the Company model be presented to the town of Nelson.

I understand that it was afterwards kept on permanent display in the Town Hall.

A couple of days before sailing for home I noticed that Chippy was busy constructing some sort of framework down on the Aft deck. My first thoughts were that he was making a bigger than normal swimming pool. After a bit of diplomatic questioning he told me it was a base for a pair of stables that were due to come aboard the next day.

By the following morning all our maintenance work was virtually complete and a notice was put at the top of the gangway saying that shore leave would finish at 1600 hrs. We had been on the New Zealand coast for something like six weeks and although we were sorry to be leaving, we were looking forward to a rest from the hectic social activities that had occupied us during our time there. They say that you can't burn the candle at both ends but we certainly gave it a good try!

Later that afternoon I was enjoying a beer out on deck when a lorry pulled up alongside the ship with a shed like structure roped down to its flatbed. Within minutes it was picked up by one of the ships derricks and swung aboard onto Chippy's prepared timber frames. As I watched a chap whom I hadn't seen before went over and opened the top half of the stable door. To my amazement a horses head appeared out of the opening and made the sort of noises that horses do when recognising someone.

Unable to contain my curiosity any longer, I went down the companionway and strolled over to the man who was by now stroking the horses head. He explained that there was another one on its way and that he and the two horses would be travelling with us to London. His job was to look after the horses during the voyage and make sure that they were in good condition for a series of show jumping events in England.

As I went to give the horses head a rub it stretched out its neck and put its chin on my shoulder and pulled me towards it.

"It seems like she's taken a liking to you, do you know anything about horses," remarked the horse minder.

"Not really. I used to do a bakers round with a horse drawn van before I left school, but the only horses I understand are those down below, all eight thousand of them," I replied.

Later that afternoon the other horse was hoisted aboard followed by a small a haystack. The horse minder was accommodated in the Pilots cabin up behind the Bridge and joined the Captain at the centre table for his meals. During the day he sat on the cargo hatch and kept the horses company and gave strict instructions that no one was to feed the horses.

However I often noticed the Baker giving them titbits from the galley early in the morning, long before the horse minder was up.

I should think that after five weeks at sea with no exercise and all the bits and pieces fed to them by various members of the crew, it would be some time before they could run let alone jump fences. Their stables were securely held down with wire cables but there was a certain amount of concern for their safety when we encountered some heavy weather in mid Atlantic. When the *Rakaia* was fully loaded she often

Homeward Bound
Cadets playing deck tennis, the box like affairs are the two stables for the horses. (Photo John Layte).

had her Aft decks awash, but Chippy's wooden base held up admirably and the beasts arrived safely in London apparently none the worse for their ordeal.

A day before docking at London's Royal Albert Dock, the Chief called me into his office to ask me what my plans were after my leave was up. I hadn't given it much thought up until then, as I assumed that Mr. Strachan the Marine superintendent would advise me on what positions would be available when he interviewed me when we docked in London.

Before I could give him an answer he told me that he would like me to return next trip and sign on as 4th Engineer. This was great news as it meant that I would be classed as a senior engineer with the added bonus of a higher salary.

We had been away for four months and five days which was an average length of time for a run to New Zealand and back. I paid off the next morning with (after deductions) 109 pound, three shillings and two pence.

I had seventeen days leave and a tidy sum of money in my pocket to see me through. One of the first things on the agenda was to visit my tailor, Phil Segal in the East India Dock Road to get measured up for a new suit.

Like most lads from London and Liverpool at that time I was what could be described as a dedicated follower of fashion. Out in Australia and New Zealand they didn't seem to bother about dress code, so when we went ashore there we had a definite advantage as far as getting the girls was concerned.

21
Promotion to Senior Engineer

My leave soon went by and I rejoined the *Rakaia* in the Royal Albert Dock in Woolwich as her new 4th Engineer. She was still being loaded with general cargo for New Zealand and was due to sail the following week.

As I lived just across the river, I was able to get home after five o'clock in the evening and return by nine in the morning whenever I wanted, with the exception of the days when I was duty engineer.

During this time the standby engineers left and the sailing ones gradually joined the ship. There were some new faces in the alleyway for the next voyage, we had a new 2nd Engineer Jack Anderson from Scotland, I had already met Jack when

Rakaia's Officers for Voyage 36 (A Double Header).
The Author is 3rd on left, back row. Brian 3d Engineer is 4th from left, back row.
Ray, Chief Lecky is 2nd from right, back row, Eddie, Chief Freezer is 3rd from right.
2nd Eng, Jack, is 2nd from right, front row and John Cowper C/Eng is 3rd from right.
Captain Ogden is centre, front row.
For some reason engineers below 4th Engineer were never included in pre voyage photographs.

he was the 3rd on the M.V. *Cornwall,* one of Federal Steam Navigation Companies ships, but effectively part of the N.Z.S.C. We became good friends over the months ahead, so I was very sorry to hear that he had crossed the bar a few years ago. I understand that he ended his days as the Marine Superintendent of a Scottish Ferry Company.

The 6th Engineer John, was a New Zealander but was naturally known as 'Kiwi'. The 5th, George was from Manchester, the 7th was from somewhere in the Home Counties and the Junior, Bob, was from the Midlands.

The 2nd Lecky Brian Anderson came from Liverpool; we became good friends and have remained so throughout our lives. The previous 2nd Lecky Dave Lang has also been a life long friend and both have contributed to this book. Unfortunately Davy has also just crossed the bar! It shows that the ship board statement of 'Board of Trade Acquaintances' doesn't always ring true.

I heard that we were also getting a new Chief Lecky by the name of Ray Pethick. He didn't arrive until the day before sailing, but his reputation preceded him, as his name was almost legendary within the Company.

We also had a new ships Doctor, Rab Davidson, he was a true Surgeon and was in his Sixties and coming up to retirement. Nevertheless he was still very much a ladies man and a great character. It was on the *Rakaia* in gale force winds that he performed what was probably his last operation before throwing in the scalpel. This was to remove the appendix from the Stewards boy who had been complaining of severe stomach pains, Rab was aided and abetted by the Mate and by Russell Birkinshaw the Chief Sparks who acted as anaesthetist by giving the boy a concoction of chloroform and ether. Russell had done many voyages on the *Rakaia* and was probably the only person who could master the intricacies of her antique Radio Shack.

The day before we were due to sail we were having our smoko in our recently soogeed down mess room when in came someone I had not seen before. He was wearing a white boiler suit that had a patina that qualified it as never having been dirty but as having been washed many times in the past. It was the mark of a typical ships electrician and this was confirmed when he introduced himself as Ray the new Chief Lecky.

He was nothing like his reputation had led me to believe, he was a quiet, smallish person with thinning combed back hair and in his late twenties. I was rather disappointed as he would have been a good replacement for Tony our eccentric ex Submariner who had now left us; in fact he was quite the reverse. In the course of our conversation he told me that this was going to be his last trip as he had recently got engaged and his future wife didn't approve of his being away for long periods. Perhaps that had something to do with his new image.

The hours leading up to sailing time were always accompanied by mixed feelings. There was an atmosphere of anticipation of new places to visit and new girls to meet, mixed with a feeling of leaving something behind and knowing that you would be away for several months.

As a new 4th Engineer there was the underlying thought of knowing that the success of the voyage rested partly on my shoulders as I would be responsible for the well being of the engine room, twice a day from four till eight o'clock.

Exactly one month after arriving home from its last voyage the *Rakaia* sailed for New Zealand. Unlike our departures from New Zealand ports where many of the local townsfolk and most of the unmarried female population turned out to wave us off, our departure from London's Royal Albert Dock went almost unnoticed. The men and in some instances women who made up the crews of Britain's Merchant Navy seemed to have a bad image among the general public in the U.K. It was never appreciated what a great debt is owed to the tens of thousands of defenceless crew members who lost their lives bringing home vital supplies during the 2nd world war.

We slipped down the Thames on the ebb tide and after dropping off the Pilot at Gravesend made our way out into the English Channel. Here, we were just one of dozens of similar ships all heading for different parts of the world. All except the Chief were on duty down below for the first couple of hours, after that we went on to the usual six hour watches until we cleared Ushant.

Once into the Atlantic we settled in to the normal sea going routine of four on and eight off. My new assistant the 7th engineer was on his first trip and it didn't seem so long ago that I was in his situation. I now found that I was playing the role of teacher, this was because he had done what was known as a student apprenticeship instead of a craftsman's one. He had spent most of his time in the classroom and had been given very little practical experience.

He was one of the new types of engineer; once he had done two years on articles he would sit and gain his 2nd Engineers ticket straight away. In some ways he was lucky to have been assigned to the *Rakaia* as he would gain experience on how to deal with situations that might never occur on more state of the art ships. Ships engine rooms were starting to become less labour intensive and I felt that it wouldn't be too far into the future before they became fully automated and the old style of 'Hands On' engineer would cease to exist.

As a 4th my main responsibilities besides being in charge of the 8 to 12 watch were to overhaul and make sure that we always had at least one set of fuel and air start valves for one generator, ready to fit at short notice. The other task that came within my remit was to control the condition of the water in the boiler by adding the appropriate amount of chemical treatment into the hotwell. To aid me in this job I had a sort of chemical set to play with.

A few days out into the Atlantic we encountered some bad weather, the ship was pitching quite heavily and true to form a bottom end on one of the generators started knocking. It happened on the 3rds watch so that by the time I went below the repairs were well underway. One of the problems with a white metal bearing running is that it throws the molten metal all round the inside of he crankcase and into the oil ways in the crankshaft. It was imperative that we retrieved all this as according to 'Murphy's Law' a piece was bound to find its way into the lubrication system and cause more trouble. This part of the operation often took longer than the fitting of new shells.

Although the 'Harlandic' geny's on the *Rakaia* were troublesome, they were easy to work on and bottom ends could be stripped out without the necessity of removing the cylinder heads. Having a good set of bearing scrapers made light work of removing excess white metal from the replacement shells. After my first trip, I always took the set that I made during my apprenticeship to sea with me.

I was not surprised when the 7th confessed to me that he had never fitted a shell bearing before so it gave me the opportunity to pass on the technique that had been taught to me by past masters of the art. The job didn't require any excessive use of the brain, but it did require a certain amount of patience and perseverance and in the case of a breakdown at sea it required an element of speed. One of the things that I remembered being told when I first started my apprenticeship was to learn how to do it right and speed would follow on behind.

5th Engineer, George, 4th Engineer, George and Junior Engineer Bob about to remove a bottom end bearing from No 6 cylinder on No Two generator. The foot of the con rod and part of the bottom end can be seen through the crankcase door. (Photo Brian Anderson).

By the end of the watch the generator was ready to run on test, the 3rd and the Sixer put it through its test routine during their watch. This was done by running the engine for one minute before stopping it and removing the crankcase door and having a good feel around to make sure there were no signs of overheating. If all was well it would be given a run of two minutes and the hands on bit would be repeated, this procedure would continue, doubling up the running time until a ten minute run had been successfully completed. It was generally considered that if everything was running O.K. by then it would be safe to put it back on service.

Other than a potentially serious incident with one of the Greasers and the odd scavenge fire we had a trouble free trip across the Atlantic to Curacao to take on bunkers. The big difference this time was that I would be in charge of the refuelling and it was my responsibility to make sure there were not any spillages.

The greaser incident occurred after he left the engine room to make the tea. He was gone for a long time so I asked the Seventh to nip up and see what had happened to him. He returned saying that the greaser was playing cards and had told him to get lost. This put me in a tricky position, I should have reported him to the Second Engineer but I would have lost any respect that I might have had with the rest of the engine room labourers.

I told the Seventh to keep an eye on things while I went up and sorted things out, however I had no idea what I was going to do when I got there. When I arrived at the Greasers Mess my greaser looked up and said.

"Off you go and tell the Second then 'Four'o.'"

I knocked the cards out of his hand, whereupon he jumped up from the table and I thought he was going to hit me. Before he reached full height I punched him straight in the side of his face and he lost his balance and went down taking his chair with him.

In general the Greasers were a tough old lot so I beat a hasty retreat back down the engine room wondering what the outcome of it all would be.

Astonishingly about five minutes later our Greaser came down carrying the teapot. He came over to me and said.

"Sorry Four'o, it won't happen again."

Whatever the rights or wrongs of my actions, I never had any more trouble from that quarter and a good working relationship was established for the rest of the voyage. However I decided not to drink the tea that watch and I advised the 7th to do the same.

In hindsight the problem was of my own making. I had sailed with this particular greaser on my first trip to sea and because I wasn't 'au-fait' with ship board etiquette within the N.Z.S.C. I probably spoke to him more than necessary. Now that I had been promoted to a senior position he probably thought he was in for an easy time owing to my previous attitude.

There was a lot of leg pulling going on and bets were already being laid that I would fill Caracas Bay with gas oil. I managed to reduce the odds by persuading Chippy to build his cofferdam a foot higher than usual and to plug up the scuppers. I tried to look calm but a sickening feeling kept hitting me in the pit of my stomach.

As soon as the fuel started to come aboard, things seemed to look after themselves, I knew all the foibles of the filling procedure, or at least I thought I did, that is until I felt the vibration and rumbling sound coming from down Aft. I quickly diverted the fuel away from the tank and hoped for the best. As the fuel came aboard it had to be distributed evenly to the various tanks so as to keep the ship on an even keel, this was done by keeping an eye on the inclinometer that was attached to the rear end of the main engine.

When the last of the tanks was nearly full I sent the 7th up to tell them to shut down the pump and then followed him up to get a sample of the fuel for the ships records. On the way I collected a couple of cans of Guinness from the Brine Room and then nonchalantly strolled out on deck quite expecting to witness a scene of major ecological proportions. To my amazement everything looked normal.

Chippy was dismantling his cofferdam, all the Gunnel Flies had gone back to their duties after settling their bets and there was no sign of the Chief. I was now able to walk down the deck with confidence and see Chippy.

"Everything O.K. then Chipps?" I asked, as I opened one of the cans of Guinness and handed it to him.

This was in the days before ring pulls, we had to use the pointed lever type of opener that had one end for cans and the other end for bottles. This was an essential part of our equipment and was carried at all times!

He took the can and after a couple of swigs answered,

"Not exactly, you lost a few gallons; let's hope you do better next time."

All hands were back on board after their swim in the beach pool so we were soon under way heading for Colon and the Panama Canal some seven hundred miles away. The next morning after I came off watch I took the engine room log book along to the Chief. We did this every day so that he could check it over and enter all the data into the official log that he kept in his cabin. Unlike the engine room version whose pages were covered in a cocktail of black finger prints, tea rings and clusters of oily spots, it was kept in pristine condition for Company use.

As I had nothing untoward to report I just dropped the log book on to his desk as I went past.

"Ah Four'o, I want a wee word with you," he called.

I thought he was going to congratulate me for only losing a couple of gallons of his fuel, but he never mentioned it.

"I've had a word with the 3rd as it looks like we might have a wee problem on our hands; he is doing a set of cards the noo, just to make sure." He said this as he studied the latest readings in the log book.

"There's nothing abnormal from the readings Chief," I remarked.

"Ah but I detected a wee change in the engine earlier this morning," he replied.

I couldn't see how he could possibly know something was wrong when I knew he hadn't visited the engine room for days. He must have read my mind as he beckoned me to follow him into his cabin, saying,

"I'll let you into a wee secret Four'o."

He stretched out on his day bed and rested his elbow on the side with his hand cupped over his ear, then got up and told me to do the same.

"D'yer hear it,"

"Hear what Chief," I replied.

"The wee difference in the exhaust notes, I think she's doon on at least two units."

As he spoke he ushered me off the day bed and exchanged places with me then placed his stethoscopic elbow in position and put the index finger of his other hand to his mouth indicating for me to be quiet. After about thirty seconds he slowly shook his head and said,

"Och she's nay reet, but we'll wait and see what the cards have to say."

When the 3rd came off watch he took the cards along to the Chief and his diagnosis was absolutely correct. No wonder he hardly ever had to visit the engine room!

The Chief went up and told the Captain that we would have to stop over at Colon to carry out repairs before crossing the Pacific. This was the last thing that he wanted to hear as it meant that we would be behind our schedule for at least three or four days. To save time it was arranged that we would be assisted by a shore gang supplied by the Americans who controlled the Canal Zone. As far as we were concerned this was good news, as it meant that we would probably get the chance to sample the delights that Colon and perhaps Panama City had to offer.

I was down below when we docked but when I went up on deck I received a shock to my system as the first thing that confronted me was a huge curtain of rich green rain forest. Before going on watch there had been no sign of land, only the vast expanse of ocean on all sides. As I took in my surroundings I realised that we were in small basin with the jungle on one side and the river on the other. A rusty railway track ran down the quay between the jungle and the ship and disappeared into the seemingly impenetrable thick green foliage of the rain forest.

There was quite a lot of activity going on down on the quay and as soon as the gangway went down half a dozen or so shore gang came aboard. They then heaved a length of canvas up over the gunnels and fed it up over the boat deck and into the engine room skylight; curiosity got the better of me so I had to go below to see what they were up to. As I entered the engine room I was just in time to see the canvas billow out into a large tube and emit lovely cool air.

While I sampled this new found luxury I noticed two of the gang manoeuvring some sort of machine into position on the top grating. This turned out to be an ice water machine complete with paper cups, I had to admit, the Yanks were certainly organised and looked after their work force. Once everything was in place the shore fitters came aboard and wanted to know where they had to start.

At first they found it hard to come to terms with taking orders from what seemed to them young boys, most of us were in our early twenties and they expected a much older crew of engineers, as was the norm in the American Mercantile Marine. However by the end of following day we had risen in their esteem, it may have been because of one particular incident that occurred, but probably because of the ice cold cans of Guinness that we gave them.

The incident in question occurred because they were not used to swinging big hammers, they were more familiar with the more state of the art 'Hydraulic Spanners' found on their more up to date ships. They were having trouble trying to undo the large nut at the bottom of a main piston rod where it fitted into the crosshead. The pair up on the grating in the crankcase were big men and must have had the combined strength of an Ox, but there was no way they could get the nut undone.

I think that was their problem, they were so big that they had little room to manoeuvre in the confines of the crankcase. One of them asked me who had done the nut up, I told him it had been boxed up by the shore gang in London, to which he replied.

"G'dam London Limey bastards."

"If you want a hand, let us know and we'll whip it off for you." I told him.

He replied by telling his mates, "This guy says he can get the nut off."

"Who's he kidding," replied another of his gang.

I thought to myself, right you bastards I'll show you what a London Limey can do.

I went up to the stores and got the 56lb hammer, then found the Junior and told him that we had better be able to undo the nut and what's more, we had to make it look easy. I tied a length of rope to the end of the spanner and while the Junior kept his weight on it I stood on the grating and gave the hammer a swing in the manner of a golfer doing a putt.

The nut shifted instantly and the Yanks couldn't believe it, but from that moment on they viewed us in a different light and treated us with a lot of respect. It's always an advantage to be on the small side if you are a marine engineer.

With the Panama climate there was a constant need to drink plenty of liquid. Consequently every lunch time and every evening we would get through several cases of beer and the empty cans were thrown out of the porthole. What we didn't know was that they floated about in the basin and slowly drifted up to one end.

One afternoon the Chief called me into his cabin and asked me if I knew how many cans had been consumed since we arrived. I told him I had no idea, but probably quite a few cases. He replied by telling me that the 'Old Man' had seen all the cans floating in the basin where they had drifted together and formed a solid metal surface. He asked the Chief Steward about it and was informed that it was all down to the engineers and he was getting concerned that we would run out of beer.

"Well Four'o, the old man reckons that 1500 cans have been consumed in the alleyway since we arrived, what d'ye have to say aboot that," he remarked.

"Well Chief, it's a bit warm down below, no one is drunk it just goes in one end and out the other," I replied.

The Chief agreed.

"That's exactly what I told the old man," he said, "I also told him that while his lot were off ashore doing their best to get a dose of clap, my lads were doon below sweating their guts oot and providing no one got drunk they could drink all they wanted."

The next morning the Chief Steward posted a notice on the board to say that beer was to be rationed to one case a week and spirits to one bottle until we reached New Zealand where we would take on fresh supplies. We drank very little spirits so the one bottle a week didn't matter; it probably affected the Mates more than it did us. However the drastic reduction in beer was a serious matter.

To make sure it was not just an excuse to reduce our consumption the 3rd managed to persuade the 2nd Steward to let him check out the stores. Sadly he reported back that the stock was right down, so we had no alternative but to accept the situation. We did manage to supplement the ration by purchasing some bottles ashore, but these were expensive and as we were still only at the beginning of the voyage we had very little money in our accounts to sub on.

On the brighter side our evenings were free to sample the delights that Colon had to offer. The bars and nightclubs were a complete new experience for me. Suffice to say and without going into lurid detail about the 'exhibitions' that were part of Colon night life, we had some good runs ashore that gave us plenty to laugh about during 'Smoko' on the mornings after.

On our Sunday off we took the train that went across the Isthmus to Panama City. It roughly followed the course of the Canal and at one point went across a long causeway across Gatun Lake. Most of the journey was through thick jungle and areas of swamp with the occasional clearing that sometimes gave us a glimpse of the Canal.

It was a single track line and the Wild West style coaches were hauled by an ancient diesel locomotive. We travelled in the rearmost coach which had an iron railed veranda at the stern end from which we had a great view of the scenery as it disappeared astern. Our day in Panama City was an eye opening experience; we had been warned not to take any valuables with us and to watch our backs. However we never encountered any trouble, even though there were plenty of unscrupulous looking characters hanging around outside the bars and back streets.

We returned to the ship by bus, which was some sort of converted lorry and not only did we share it with a motley crowd of locals, there was a group of wild looking natives who had with them a pair of small pigs and several chickens that were given the freedom of the bus. After rounding up their stock they departed the bus just before Darien about half way across the Isthmus and disappeared into the forest.

The day before we were due to sail, I was leaning on the gunnels and enjoying a can of ice cold lager, while taking in some of the jungle atmosphere. Suddenly I noticed a green line painted along the top of one of the rusty railway lines that went down the quay. I was sure the green line was moving; perhaps the old man was right about our drinking after all!

I went down the gangway and across the quay towards the track and the closer I got, the more it seemed to move. I began to feel a bit concerned until on close inspection it became apparent what it was all about. A dense column of the largest ants that I had ever seen were marching along the top of one of the rails and on their backs they carried a large piece of green leaf that waved about as they went on their way. They were so densely packed together they appeared to be a solid green line. I took a swig from the can of lager that I still had with me and felt quite relieved that I hadn't been seeing things.

I followed the line into the jungle for about fifty yards until it suddenly disappeared into impenetrable undergrowth and all the while I was conscious of a low buzzing noise which seemed to be coming from the ants. As I listened I also became aware of a rustling sound that was coming from the thick layer of dead vegetation beneath my feet, suddenly a huge claw came up out of the ground. It was definitely time to leave!

When I got back aboard I mentioned my little expedition to one of the shore gang, who gave me some good advice.

"Gee Mack, you don't want to go into the jungle around

here, it's full of snakes and all sorts of things that sting or bite. The claw you saw belonged to one of the giant land crabs that live in burrows all around here; the place is infested with them. Another thing, you want to keep your wits about you coming back from town at night. The locals are crazy drivers and if you got yourself knocked down by a hit and run, the crabs would be on you and pick you clean before the ambulance could get to you."

He went on to tell me that the reason why the grass was kept so short all around the Canal Zone administrative buildings and infrastructure, was to help keep all the creepy crawlies at bay. I had noticed this on my previous transits and had assumed that it was just for decorative purposes. It was always a rich green and would have made any golf club green keeper quite envious.

I mentioned my expedition into the jungle and my encounter with the crabs to the lads during 'Smoko' and Kiwi the 6th Engineer insisted that I showed him the place. I knew he was into that sort of thing as he took a great interest in the wild life that we sometimes encountered on our travels.

One of the ghoulish stories concerning the building of the railway the yanks told me about was the pickling of human cadavers in barrels for shipment to various countries for the purpose of medical research. The death toll for the first seven miles of track ran into thousands and due to the swampy terrain it was difficult to dispose of the dead. They came up with the idea of selling them for medical research, so full cargoes of pickled cadavers were sold and the proceeds were used to finance the local hospital.

One of the benefits of our stop over at Colon was that I was able to purchase from the Commissary of the Canal Zone some superb quality work gear in the way of shirts and trousers. These items were issued to the American work force of the Canal and were impossible to obtain any where else.

22
Panamanian Quickstep

After a stay of five days we said goodbye to our American engineers and entered the locks for our canal transit. We were a day out into the Pacific when a huge commotion erupted from the junior's cabin. When I looked in he was standing on his day bed and throwing anything he could lay his hands on in the direction of the open door of his wardrobe.

"There's a bloody great monster of some sort in there, it nearly took my hand off." He screamed.

After taking a cautious look, I understood why he was in such a state, the biggest crab that I had ever seen was sitting in there and waving a huge claw in my direction. As no one was prepared to remove it, the creature would have to remain where it was until a solution could be found. When Kiwi came off watch and was told about it a huge grin appeared on his face, which put him in the frame for having put it in there.

The junior told Kiwi to get rid of it otherwise he was going to fetch the long crowbar from the stores and smash it up. To

Engineers and ships Cook (seated) enjoying the crab sandwiches.

keep the peace Kiwi lifted it out of the wardrobe; the beast was so big it required both hands to hold it. There was no point in throwing it overboard, as being a land crab it couldn't survive in the sea. Someone came up with the brilliant idea of taking it along to the galley to see if Gerry the Cook would cook and dress it for us. When he saw it he said, "I know the crew sometimes get crabs, but you engineers must have sore crutches if you breed em like that."

The poor creature ended up in sandwiches of freshly baked bread and went down very well with a couple of cans of Guinness from our depleted ration.

As it turned out, it wasn't such a brilliant idea after all, as the next day we all had a severe dose of 'Panamanian Quickstep' that necessitated keeping within range of the loo's for the next twenty four hours and resulted some record breaking exits from the engine room for those on watch.

Our voyage continued towards Samoa and was comparatively uneventful until a day out from Apia we received a bad weather report. Owing to the bad anchorage in the lagoon the Captain decided to by-pass it and go on to Suva.

It was at this point that the 'Phantom Crapper' manifested himself. Brian the 3rd went down Aft for a spot of morning bronzing on top of the spud locker and was confronted by a large turd with a cocktail stick in it that had a flag on it. Close inspection revealed a notice written on the flag that proclaimed 'The Phantom Crapper has struck'. Over the following months we continued to find similar deposits with the note on the flag saying, 'The Phantom Crapper Strikes Again and will never be caught'. They were invariably found in places that were known to be frequented by the engineers so we took it personally.

New sightings occurred every month or so, until in the end we posted a notice on the board to say that in his absence 'The Phantom Crapper' has been sentenced to a fate worse than death. We laid several traps for him over the duration of the voyage, but to no avail. Before we left New Zealand for the final time a notice appeared on the board made up of letters cut out of various publications informing us that he was going to strike in the Chief Freezers cabin at 1500hrs on the coming Saturday.

By then everyone was under suspicion, so by being in the Chief Freezers cabin at 1500hrs you made sure that you were no longer on the list of suspects. Unfortunately this meant that we would have to forgo our Saturday afternoon session ashore. By 1400hrs the cabin was crammed full, everyone had bought along a case of beer and they were stacked up just inside the door. During the next hour several new and ingenious ideas on what we were going to do to The Phantom were discussed and as 3 pm approached the pile of cases was getting dangerously low. The cigarette smoke was so thick you could cut it with a knife and the empty cans were ankle deep underfoot.

It was generally agreed that there was no way the 'Crapper' could escape from this one and as the minute hand of the clock on the Chief Freezers bulkhead came up to 3.00 pm, all eyes were on it. Suddenly someone shouted, 'Look' and pointed towards the wash hand basin. There suspended and slowly revolving on the end of a length of string was a well preserved turd with the now familiar flag and notice. While we had all been watching the clock the 'Phantom Crapper' had lowered it down through the ventilator from the boat deck above.

There was a huge rush to get out of the cabin but the pile of empty cases and the cans underfoot caused a log jam of monumental proportions at the door. By the time someone eventually reached the boat deck the 'Phantom' was well and truly gone. For some reason that was the last time he struck and we never had any more sightings for the rest of the voyage.

Because we by-passed Apia, Ray the Chief Lecky, showed his true colours. This was his last trip and as he had never visited Samoa, he was looking forward to a session ashore. As I went on watch that evening, I passed his cabin and saw that he was having a drink with the Chief Freezer. They were on day work so they could afford to have a few drinks in the evening as they had the rest of the night to sleep it off. I looked in and jokingly said.

"Never mind Ray, there's not much to do in Apia; you can make up for it when we get to Suva."

My watch was uneventful and when I came up at midnight and stepped into the alleyway, Ray shouted from his cabin.

"Come and have a drink Davy boy."

I looked in on my way to the wash room and couldn't believe my eyes. The cabin looked as though a bomb had gone off; the place was an absolute disaster area and stank worse than a cess pit.

During the evening they had been joined by the 2nd lecky, the 2nd Freezer and Gerry the Cook. They were all paralytic and someone had spewed up in the wash hand basin and half up the bulkhead. The floor was littered with empty cans which far exceeded their weekly ration and there were numerous bottles of rum and whisky rolling about. Ray was the last man standing, the rest of them were all out for the count and in the most uncomfortable positions imaginable. I wisely declined Rays offer, got scrubbed up and joined the 7th for a quiet beer before turning in.

The next morning our Steward understandably refused to enter Ray's cabin to clean it up and duly reported the fact to the 2nd Steward. Shortly after, Ray was visited by a deputation made up of the Chief Steward, the 2nd Steward and the Mate. He refused to let them enter his cabin unless it was to join him for a drink. For some reason they agreed; 'For diplomatic purposes' I was told later, but when I went back on watch at 8 am they were all still in there.

Somehow the situation resolved itself and when I looked in on Ray before lunch his cabin was as clean as a new pin. The incident was soon known about throughout the ship and enhanced his dubious reputation no end. By the time we docked in Suva things had pretty well returned to normal, but before our visit was over, I was once again to witness Ray's eccentric behaviour.

He accompanied Brian and myself to our old haunt 'The Golden Dragon'. It was there that I experienced his Jeckle and Hyde type of character for the first time. Once he had downed a couple of whiskies he became a different person. The warning sign that we were about to get into trouble was when he produced a pair of tight fitting leather gloves from somewhere about his person. After putting them on he always punched the fist of one hand into the palm of the other and said, "Right I'm going to sort out that big bastard over there."

Ray was only about five foot nine and around ten stone in weight and would never have stood a chance if his prospective opponent had turned nasty. Luckily on that occasion everything went well and ended up with 'The Big Bastard' buying Ray a whisky and the pair of them doing some sort of dance on top of the bar. Brian reckoned it was Ray's perverse way of getting free drinks. However he was not always so lucky and often ended up with a bloody nose.

One morning sometime later while we were in Sydney I happened to be on deck prior to going in for breakfast. Suddenly a police Black Maria careered around the corner of the sheds and screeched to a halt at the bottom of the gangway. The rear doors burst open and out tumbled Ray wearing only his Y-fronts, the doors slammed shut and the vehicle did a U-turn and sped off in the direction whence it came.

Ray climbed the gangway at lighting speed and disappeared into the accommodation leaving the Cadet on gangway duty with his jaw wide open and a look of shock horror on his face. I doubled back and found Ray in the wash room having a shower and I could see that he was covered in bruises and looked in a proper mess.

"Whatever happened to you this time Ray?" I asked.

"I'll tell you about it later Davey boy, I must get cleaned up and get some breakfast, I'm starving," he replied.

During 'Smoko' he gave us his account of what had happened to him. Apparently he had been invited to a party ashore; this was a silly thing to do as we never went to these sort of invitations alone. He had taken a case of beer and a bottle of spirits along with him but he had no recollection of where it was. The next thing he remembered was waking up in the early hours of the morning under a hedge in the front garden of a bungalow wearing only his Y-fronts. He had no idea of where he was, so the only thing he could do was to knock on the front door of the bungalow.

It was opened by a woman wearing a see through night dress, as he stood there mesmerised trying to come to terms with the situation, the woman's husband or boyfriend appeared in his Y-fronts and punched Ray on the nose and called him a bloody pervert. In the mean time the woman had called the police. They soon arrived and carted him off, but before he managed to explain to them what had happened they gave him a good going over in the back of the van. Eventually he managed to convince them that he was not some sort of pervert, but a distressed British Seaman.

Soon after our arrival at Auckland in New Zealand, we received two pieces of unwelcome news. The first concerned Brian our Third Engineer. Our ships agents had informed the Captain that Brian had to return to England straight away. I was ashore at the time and when I returned to the ship he had already packed his gear and left for home. I never found out all the facts, but I understood that there was some sort of serious family problem. The other piece of news was that we were going to load a cargo for the M.A.N.Z. run.

The M.A.N.Z. line stood for Montreal, Australia, New Zealand and was formed in 1936 as a co-operative arrangement between The Bucknall Steamship Company (Ellermans), the Commonwealth & Dominium Line (Port Line) and the New Zealand Shipping Company. The purpose of the Line was to manage Canadian shipping services to and from those countries.

What it meant for us was that after leaving New Zealand we would go across the Tasman Sea to Australia then across the Pacific through the Panama Canal then up the Eastern freeboard of America, stopping at Galveston, Philadelphia, New York then back down to Bermuda. After Bermuda we were to go up to Boston before going on to Canada, then across to Newfoundland for final loading of a cargo for New Zealand. We would then load a frozen cargo for home. This was effectively two trips in one and was known as a 'Double Header', these sort of voyages usually lasted for about ten months.

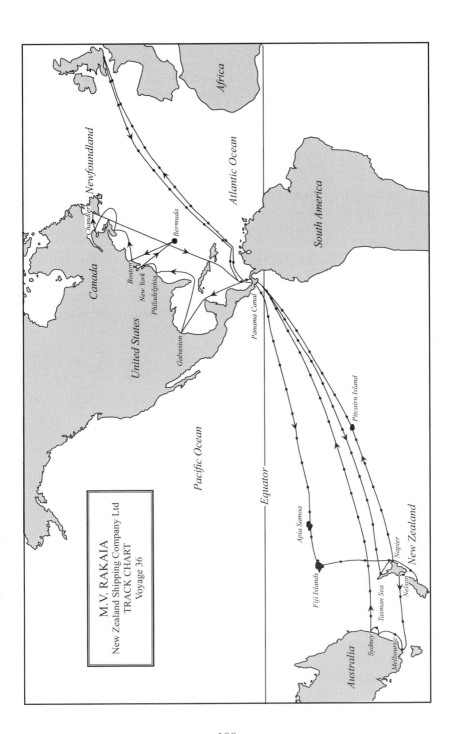

M.V. RAKAIA
New Zealand Shipping Company Ltd
TRACK CHART
Voyage 36

As Brian had unexpectedly left us we were left without a Third Engineer and I was sorry to see him go as we were good shipmates and had many good times together. The day after Brian departed the Chief called me into his office and said, "Ah Four-o, a wee word with you, as you know we have lost our Third, I'd like you to take his place. I know you are conversant with the responsibilities involved and it's always better to have the devil you know. I've got a replacement fourth flying out to join us. In the mean time the 2nd will re organize the watches until he arrives. When you knock off you can move your gear into the Thirds cabin and get sorted."

This was indeed good news to me, not only did it mean an increase in wages from £56 to £69 a month it also meant that I would be back on the 12-4 watch when at sea, which was the one I preferred and be relieved of the bunkering duties. It also had the advantage of a slightly larger cabin, the only drawback being, I would have to be the duty engineer on docking days.

Our new Fourth, George, came from Liverpool and it soon became evident that he knew his job; he was great company and proved to be a good shipmate for the rest of the voyage. Having been made up to Third I was required to have two new sets of epaulettes, these would normally have been purchased from the Slop Chest, but unfortunately the Chief Steward informed me that there were none in stock. This meant that I would have to continue to wear my old ones.

One day while at dinner the Captain noticed that I was still wearing a Fourth's insignia and took me to task on the matter saying.

"Tell me Mr. Carpenter, why do we seem to have two Fourth Engineers aboard my ship?"

I explained that there was no gold braid in the Slop Chest and that in view of the situation the Chief Steward had ordered some, but it would be some time before he received the order.

"At times like this you have to be resourceful and use your initiative," he boomed across to me.

The next morning at breakfast he came over and gave me a paper bag containing an assortment of braid and said, "See what you can do with that."

I set too with needle and cotton and managed to do a passable job on my tropical set so that at dinner that evening the Old Man had no cause for complaint, he just looked across at me and nodded with the faint hint of a smile.

Forty five years later while attending a 'Durham Association' Ladies Luncheon at Liverpool's Maritime Museum I had the pleasure of meeting him again. Although in his nineties, when I mentioned the epaulette incident, he laughed and said,

"Yes I remember it well."

I am sure that he did, as he was a stickler for dress code in the dining saloon.

The Durham Association was formed in 1958 by a group of Cadets that stayed on board the M.V. *Durham* for the current full thirteen month voyage. Six months of which was spent in Galveston Texas while she underwent major engine repairs after she had broke both crankshafts of her Sulzer main engines. The Association has since developed into the main social contact for ex-employees of the New Zealand Shipping Co & Federal Steam Navigation Company. It holds lunch meetings in New Zealand, England and Australia and distributes a news letter to its members several times a year.

After discharging most of our general cargo we went down the coast to Napier to start taking on our frozen cargo. This was one of our favourite ports as we were always made welcome by its population of beautiful girls. The harbour sits beneath a very high cliff face known as the Bluff, this was one of several places, where, would be marine graffiti artists could exploit their talents. Under cover of darkness, crews of visiting ships seemed to compete to paint the name of their ship on the most inaccessible parts of the cliff. The *Rakaia* had a distinct advantage with this type of artwork as she carried a crew of very fit Cadets who couldn't wait to risk life and limb to show off their talents.

As the *Rakaia* was a regular visitor to Napier her name only had to be given a fresh coat of paint and the date changed to coincide with her latest visit. It's around forty years since she and many other regular visiting ships docked there, so I would imagine that very little remains of that part of British marine history, the severe storms that sometimes lash that part of the Pacific Coast would have seen to that.

I was on deck when we departed Napier and as we pulled away from the jetty I could see that someone had painted in bright red letters on the timbers between the piles 'RAKAIA 63'. I also noticed Frank the Bosun up on Fo'castle giving a group of Cadets a mouthful of colourful language that centred on his stock of red paint while at the same time pointing from the paint locker to the receding jetty.

This incident reminded me of something that happened while we were berthed in London's Royal Albert Dock. Moored up on the opposite side of the dock was a ship by the name of *Alfred Janus*, the next morning the Alfred had acquired an S on the end and the J had been painted out.

I heard that the old man was not very pleased when it came to his notice and he called all the Cadets together and gave them a good talking to.

Someone said that as he left them to consider the consequences he had some difficulty in disguising a smile.

A view from the top of Napier's Bluff showing the *Rakaia* with a Watts Watts & Co ship berthed alongside. (Photo John Layte).

I seem to recall that the ship in question was of German registry.

After Napier we went round to Wellington which is at the bottom of the North Island in the Cook Straight. It's an area known for its strong winds therefore Wellington was known as 'Windy Welly'. They say you can tell when a man comes from there as he always leans forward at an angle of forty five degrees with his hand on his hat. While in Wellington we took on fuel from a small oil tanker, it came alongside while we loaded our last consignment of frozen lamb before we sailed for Australia.

Although it was summer time in these waters we left Wellington in heavily leaden skies and encountered heavy seas as we left the Cook Straight and entered the Tasman Sea. This was not unusual as it was an area well known for its rough weather.

Our first port of call was to be Melbourne, some 1250 miles away, the weather deteriorated very quickly and by the end of the second day we were encountering very large seas. All the dead lights and watertight doors were battened down with news of worse to come.

At first I just thought it was going to be another bit of rough weather, but during the night on my 12 to 4 watch things started to get serious. The Chief phoned down to tell me to stand by the controls and watch the revs, he added that a couple of the lads were coming down to assist me.

Wellington Harbour
The *Rakaia* is berthed at bottom right alongside the floating drydock. Berthed on her Port side is one of Ellermans City Boats. At the far end of the quay are N.Z.S.C.'s *Rangitoto* & *Rangitane*. Ahead of them is one of Federal Steam Navigation Company's ships. Just leaving for London via Panama is Shaw Savill's 26,463 tons gross *Dominium Monarch*. (Photo Paul Wood).

At times we were rolling so badly that it seemed she wouldn't come up again. On one particular roll I watched the inclinometer on the after end of the main engine go over to 45 degrees and realised that I was mentally urging it to come back up. We were now in a dangerous situation; the engine governor could not cope with the sudden change in load on the propeller. It was coming out of the water at the top of each huge wave and then deeply burying itself after we surfed down into the next trough.

We had to take turns on the fuel lever to counteract the huge change in revs. This was done by constantly watching the tachometer and at the same time anticipating when the revs were going to need correcting. This was achieved more by the feel of the ship under your feet; this combined with the severe vibration you soon got the knack of judging when to open or shut the fuel lever before the revs reached their critical point.

After fifteen to twenty minutes the control would be handed to someone else as it soon became difficult to concentrate while trying to keep your balance. It was necessary to use both hands on the fuel lever while at the same time looking up at the large tachometer which was a good twelve to eighteen inches above eye level. A Gunnel Fly would have thought we were praying, I know from my own feelings at the time it wasn't possible to think of anything except to try and keep the pointer on the tacho within the bounds of safety. However I have no doubt at all that we all offered up a prayer at one time or another during those hours of need!

The Fourth went down the shaft tunnel to check things out and came back with the news that the stern gland was leaking badly, he returned with the Seventh to see if they could tighten the gland nuts and stem the leak. They came back after about half an hour, both soaked to the skin and reported that the gland was in as far as it would go and water was still pouring in. We put the bilge pump onto the tunnel well and hoped for the best.

The Second phoned up to the Chief and informed him of the situation, whereupon the Chief replied that he would be closing the water tight door to the shaft tunnel in five minutes. This was done from a large hand wheel just inside the engine room door. It would give us time to make sure no one was left in there. Had any one been left down there they would have been able to use the tunnel escape ladder which would have taken them up to the lazarette. With the after deck constantly awash they would almost certainly been swept overboard before they reached the accommodation. This was the only time I experienced a watertight door to the shaft tunnel being closed in earnest.

Not only did we have trouble keeping the oil pressure up to the main engine, we occasionally lost suction on the sea water pump. This could only be due to the ship exposing the sea water inlet which was way down below the turn of her bilge.

At one point while I was in praying mode at the controls, I heard a terrific crash from somewhere behind me. I managed a quick look round and was astounded to see that the large steel work bench had slid across the plates and smashed into the side of the main engine cooler. It had always been assumed that this extremely heavy piece of equipment had been bolted down.

I had been in some quite rough seas since joining the *Rakaia* and it had never moved a fraction. Eddie the Chief Freezer had done several trips in her over the years and confirmed that the bench had always been there and had never moved. Before it could be secured it started to slide back towards its normal position, someone managed to get a lashing over the vice that was bolted on one end, but this made it skew round and crash into and fracture the fuel pipe to number one generator.

Fortunately the fuel leak wasn't serious, but by the time it had been temporally patched up, the surrounding plates were covered with a fine film of gas oil. It was impossible to walk in this area until the Greaser managed to clean up it up sometime later. In the meantime the only way to traverse it was to wait for the ship to roll in the direction that you wanted to go, then emulate a surfer and glide across.

Standing at the controls I was not in the firing line of the fine jet of gas oil that came from the fractured pipe. However the Second and the Fiver who made a temporary repair to the pipe, both received a good soaking and took turns to go up and have a shower and something to eat so that we could maintain having four engineers on watch.

This was the only time while I was at sea that we didn't have regular cooked meals. It was impossible for Gerry the Cook and his intrepid staff to prepare meals in the normal way. The galley was no place for the faint hearted when we occasionally encountered bad weather, but with the exceptional conditions prevailing at the time it was positively dangerous to attempt to cook on the griddles. However the Baker still managed to make adequate supplies of bread and other interesting and tasty bits and pieces so no one went without, that is except the odd one or two who were suffering from the old 'Mal de Mer'.

When things calmed down I had a look at the Seconds bit of pipe repair work. He had wrapped a length of steam packing around the pipe in a continuous spiral and pulled it tight, he then cut a Guinness can open and wrapped it round the packing then secured it with four Jubilee Clips. It never leaked a drop and was still in place when I left the ship at the end of the voyage. It was yet another example of the saying that there is nothing so permanent as a temporary repair.

We were lucky that for some reason our generators never gave us any trouble during our crossing of the Tasman Sea. After twelve hours of fighting to control the engine revs we started to notice the ship was behaving in a much kinder way. The bridge phoned down and said that the wind had gone down to storm ten and things up top were improving by the hour. By the time dawn broke we were up to full engine revs and back on course for Melbourne and back on normal watches.

The ship had sustained some damage to various hand rails and to one of our Whalers up on the boat deck. The worst damage was to the winches up Forward. One had been hit by a heavy sea and two of its holding down bolts had snapped and the rest were found to be bent. All the winches up Forward had suffered water ingress to their control boxes and gave the Leckys plenty to keep them busy for the following few days.

I was also given an extra workload, this was because when the workbench collided with the cooler, it shot both its heavy steel drawers out on to the plates together with their contents of fuel valve spares, most of which ended up in the bilges. The new parts came coated and sealed after being dipped in hot wax, so they never suffered. My stock of used, but serviceable spares, plus the nozzles that had been painstakingly overhauled were in a sorry state. By the time many of them had been salvaged they were well past their usable condition. The rest probably rattled around in the bilges until the *Rakaia* went to the breakers yard nine years later.

Although we were through the worst of the bad weather, which the Captain recorded as a Typhoon, we still continued to ship the odd rogue wave and the sky remained overcast. Normal meals were resumed although the table cloths were kept damped down and the table sides were kept up to stop our plates from sliding off for the rest of the crossing to Melbourne.

Compared to New Zealand our shore side activities were somewhat subdued, we did have a football match with a

Heavy seas continued until we reached Melbourne.

team from the Royal Australian Navy, but the less said about that the better. Our Chief was a keen soccer fan and after returning from a shore visit, declared that we were going to play the Australian Navy the following afternoon.

This was bad news as it meant giving up our Saturday afternoon on the town. Somehow a full team was conscripted from among the deck and engineer officers, but as none of us had played football for years we were a complete shambles and were beaten by an unmentionable amount of goals.

However the 'après' massacre drinks session turned the tables on our Colonial colleagues and we regained their respect by being able to walk out of the pub in the early hours of Sunday morning, while leaving them incapable and at the mercy of their Shore Patrol.

From Melbourne we went back into the Tasman Sea and up to Port Kembla. Although this was the centre of the Australian steel production and the area was very industrialised it had one of the best beaches that I had ever visited. The beach and its backdrop of huge sand dunes curved away and disappeared into the distance. It also had some very heavy surf, probably caused by the 'Typhoon' that we had recently survived.

We stayed for a few days until the Australian Wharfies eventually agreed terms with our agents and got on with the job of loading large rolls of steel sheet destined for Canada. We then got underway for the short trip up the coast to Sydney to take on our final consignment of cargo before leaving for Panama.

23
Close Call

Our voyage across the Pacific was memorable owing to a potentially dangerous incident. It occurred a few hours after transferring some of the fuel that we had taken on in New Zealand up to the settling tank and running it down to the purifiers then back up to the daily service tank.

The trouble first manifested itself during my watch with a slight drop in revs on the main engine, a quick check on all the temperatures and pressures showed that everything was normal, so we thought that we were probably going against a strong tide. That all changed when the 2nd Mate phoned down to ask us if we had a scavenge fire, as we were making black smoke.

I sent the Sixer up to check it out, he returned saying,

"It's a bright moonlight night up there, but everything astern is pitch black with smoke, it's definitely not a scavenge fire though."

I phoned the Chief and explained what was happening and he came to the conclusion that it was probably to do with the fuel that we had taken on board in New Zealand and we were now using.

One of the Junior engineers jobs every watch was to open the large valve at the bottom of the settling tank and drain off the accumulated water and sludge. Even under normal circumstances it was amazing how much came out each watch, but it was now draining a seriously abnormal amount. The centrifugal filters were also packed solid before their regular change over for cleaning. It was decided to transfer the bad fuel back down to the double bottom tank and keep it for the boiler when in port.

During the 2nds early morning 4 to 8 watch the main engine slowed down and stopped. The alarm went off and we all rushed down below, as I have already mentioned this was something that did occasionally happen with Bermiester and Wain double acting engines so we were not unduly concerned.

However the Chief was not too happy about things, after we bled the fuel system he told the 2nd to take the bottom indicator cocks and me to take the top ones. Everyone else was told to clear the engine room and wait by the mess room just inside the deck entrance. The Chief took the controls and phoned the bridge to say that we were ready to start.

I heard the telegraph ring and the next thing I knew was the ship seemed to rock with a huge explosion. I was already running along the tops starting to shut off the indicator cocks which were shooting long jets of flame up towards the engine room skylights. There seemed to be a secondary explosion a split second after the first, which came from somewhere above in the region of the boiler.

It all happened so fast I never had time to think, I didn't look up, I just kept running until I reached the after end of the main engine and then carried on into the short alleyway that led out on to the deck. As I dived amongst the rest of the lads who were waiting there, as instructed by the Chief, I heard a lot of crashing and banging going on behind me. I heard one of them say, "Jesus Christ 'Three O', we thought you were a 'gonna' there for a minute."

When I looked back the engine room was full of black smoke with snowflake like pieces of asbestos drifting down through it. Our immediate concern was for the Chief and the Second who were still somewhere below. There wasn't time to discuss what to do, I rushed into the adjacent mess room and grabbed some rags. Unfortunately there wasn't any water available, but there were several full cans of Guinness on the table so I quickly opened them and poured the contents over the rags. I gave half to the Fourth and told the Fiver to breakout the horrendous asbestos fire suit and follow us down below and then told the rest of the lads to stand by in case we needed fire fighting equipment.

We wrapped the rags around our heads and felt our way to the first set of steps, it was much clearer by the time we reached the lower set of indicator cocks, but there was no sign of the Second. I looked down over the handrail and was relieved to see the Chief on the phone and the Second heading for the ladder. As we met at the top, he pulled a face.

"Bloody hell" he said, "you two smell like a brewery, have you been on the piss again."

"No but I've just wasted some good Guinness coming down here to look for you," I replied.

We all smiled and headed up the ladder to be confronted by the Fiver encased in the asbestos suit on his way down. We threw our Guinness soaked rags at him, squeezed by and left him to get in some fire drill practice.

"I think we may have a wee spot of bother," remarked the Chief as he looked up towards the boiler when he joined us on the engine tops, which together with the surrounding area had acquired a mixed coating of asbestos and carbon dust.

He added, "The bridge were shitting themselves, they phoned doon, not expecting an answer. Apparently a wee flame shot out of the funnel some fifty feet high and lit up sea for miles and a shock wave went out from around the ship. They were under the impression that the engine room had blown up."

He told the Fourth to go up and check out the boiler while the rest of us got on with removing all the fuel valves. By now we had been joined by the Freezers together with the Lecky's, plus the engine room labourers so there were plenty of helping hands.

The smoke had now cleared sufficiently to see that there was a gapping hole in the front of the boiler where the furnace front should be, it was now lodged some fifteen feet below on top of the exhaust piston yoke of number six unit. As it had blown out it had taken a large area of lagging with it, fortunately the fuel supply connection sheared off beyond the valve which undoubtedly saved us from a serious engine room fire.

The piston yoke was a huge steel casting, so it suffered no damage. While the labourers, under the supervision of Ray, the Chief Lecky got on with the job of lifting the furnace front off it, the rest of us got on with the job of removing the thirty two fuel valves from the main engine.

The problem was soon revealed when we found that one of the valves had part of its nozzle missing. This had allowed the combustion pressure to get into the fuel system and stop the engine. When we bled the fuel system it also allowed the fuel oil to pour into the combustion spaces culminating in the big bang that required the watch up on the bridge to seek a clean set of underpants.

The Chief was concerned as to the whereabouts of the missing piece of fuel nozzle. It may have been blown out of the exhaust ports; on the other hand it may have been still sitting on top of the piston waiting to do severe damage to the pistons and the cylinder liner. We fished about through the valve apertures using a small magnet that was screwed to the end of a rod but had no luck! As the faulty valve was in the lower combustion chamber we would have to strip out both the top exhaust and the main piston to make sure the broken piece wasn't in there. With the ship rolling about this would have been a difficult and dangerous job.

It was decided that the best thing to do was for someone to crawl through the bottom exhaust manifold and check out the combustion space through the exhaust ports. Kiwi the 'Sixer' drew the short straw, so while he wrapped rags around his knees and elbows we unbolted the cover plate from the end of the manifold and engaged the turning gear and hung the 'Turning Gear In' notice on the controls.

This was an early sort of 'Health & Safety' procedure that was always done when working on or in the main engine. In our present situation, with the ship laying beam on to the seas it was quite possible that a wave could give the propeller a bit of a turn. If Kiwi happened to have his arm through the exhaust ports at the time it would have been neatly sliced off.

After we tied a length of rope to each ankle he disappeared into the manifold clutching his torch in one hand and the magnet in the other. To everyone's relief, he crawled out backwards after an agonising ten minutes looking like he had been liquidized, but brandishing the piece of nozzle on the magnet.

While Kiwi was doing his bit! We fitted a temporary cover plate over the gaping hole where the burner had once been and fitted a reconditioned set of fuel valves to the main engine, then bled the fuel system in readiness to get underway. We also checked out the boiler exhaust heating flap valve, this had been shut when the engine had gassed up, but for some reason it couldn't have closed properly. This had allowed some of the explosion to divert through the boiler and cause the damage. There wasn't much we could do in our present predicament, except to work the operating ratchet back and forth a few times in the hope of rectifying it.

We had to take turns doing it as the temperature up behind the boiler was in the region of a hundred and forty degrees. The only ones that were exempt from this were Kiwi; he was still trying to hydrate himself, giving us concern as to the possibility of a spell of water rationing. The other was the Chief; this was on the grounds that the soles of his carpet slippers might melt down. To the amazement of the Bridge, we were under way in just under two hours. We had been very lucky to have got away with it so lightly.

We arrived at Galveston without any further major incidents and once we were tied up alongside the quay the Donkey man flashed up the boiler which we had repaired on route. As I was now the Third I had to forgo the joys of docking day and stay aboard as duty engineer and watch the rest of the lads hotfoot it into town.

After lunch and with everything behaving itself down in the engine room I joined the Chief in the smoke room to watch a film on the television that had been installed for the duration of our visit. Suddenly, the Fourth together with Ray, the Chief Lecky burst in.

"Quick! You had better come and have a look outside," shouted Ray.

By now the Chief was fast asleep with his mouth wide open and acting as a perfect fly trap so I left him to get on with it and went out on deck. The whole waterfront was enveloped in a dense cloud of oily smelling smoke. It was pouring out of our funnel and rolling down the sides in an almost liquid state. I rushed down the engine room and up to the boiler flat and shut down the boiler.

There were heavy fines for smoke pollution in the American ports and we were now in line to get first prize. The problem was caused by the bad fuel that we had taken on back in New Zealand; we were now running the boiler on it, hoping to get rid of it. As it was not even fit for that, it had to be pumped out into a waste oil barge, we then refilled the tank and topped up the others from a road tanker that came alongside on the quay.

The initial use of the bad fuel in the main engine was probably what caused the fuel nozzle to give way and potentially put all our lives at risk. The fuel valves were precision pieces of machinery, weighing in the region of forty pounds. During overhaul, after the seats had been lapped in to perfection, all the parts were washed in Solar Oil and left to dry. The use of any sort of rag at this point would leave micro pieces on the surfaces and give cause for trouble.

Being a Sunday, the dock area was practically deserted so luckily the incident wasn't reported to the powers that be. However the Cadets had a busy day on the Monday soogeeing down the funnel and upper superstructure then giving it all a coat of paint. There was also a secondary problem caused by the liquid smoke. It had settled on the water around the ship giving the surface a psychedelic rainbow effect which made it look as though we had been pumping out our engine room bilges. This of course was another serious offence. Fortunately a stiff breeze sprang up and disposed of it in the direction of an American Lykes Line vessel tied up further down the quay.

This was the only time I saw the Chief get really angry, his normal shows of anger were somewhat tongue in cheek, especially so at the dinner table where he liked to put one over on the 'Old Man'. On this occasion, after an abusive tirade about the ancestors of the oil barge skipper back in New Zealand, he stormed into his cabin to concoct a strongly worded letter to the oil company, whether anything came of it I never found out.

We stayed in Galveston for several days and enjoyed every minute of it, that is except for a few hours one evening where I and the Fourth did time in the local jail. America has some strange customs; one of them seemed to be that we weren't allowed to chat up local girls. We were just talking to a couple of beautiful Mexican girls, when a Texas Rangers patrol car pulled up alongside us and out got two Rangers brandishing rifles. They ushered us into the back of the car and carried us of to jail where we spent a couple of miserable hours. After some questioning about our identity they let us go, saying they had pulled us in for our own protection. It was my first experience of being part of the American segregation system and it made me feel angry.

One evening Brian (The Bear) and I inadvertently gate crashed a private function. It all happened because we asked the cab driver if he knew of anywhere that was holding a dance. He said that he had read in the local newspaper that there was some sort of do going to take place at the Hotel Galvez.

"That sounds fine, could you drop us off there then," asked Brian.

The hotel turned out to be a very imposing building that overlooked the beach. It had a colonnaded entrance and a large marble effect foyer containing an ornate water feature. The whole establishment reflected its deep Southern origins of opulent splendour.

It all seemed very quiet inside with no signs of any girls or dance taking place. We presented ourselves at the reception area and Brian asked as to the whereabouts of the dance. We were both wearing our tailor made Italian cut suits, which together with Brian's Scouse accent and my Cockney one, must have struck the receptionist as a little strange because she asked us if we were members. "Of course," I said for no particular reason, as I started to put my hand into my inside pocket for my fictitious membership card. Before I had a chance to say that I had left it in the car and probably because she couldn't understand a word we spoke, she smiled and said,

"Yo' all go right on down the corridor; the function room is on the left."

Brian raised his eyebrows at me and we strolled off in search of the function room. We entered to the sound of a small dance band playing a waltz with two or three couples gliding around the floor. Not exactly our cup of tea! However we decided to give it a go and see if things livened up, we bought a couple of beers at great expense and sat down to weigh things up.

Things didn't improve so we thought about moving on, there were only about thirty people in the hall and these were all sitting at tables down at the far end.

"Come on Brian, let's go and ask a couple of the girls down there for a dance."

"No way, they are all with fella's, they'll just tell us to get on our bikes," Brian replied.

"Well we are leaving anyway, so we've nothing to lose, come on lets give it a go," I said, as I got up and started the long walk down the hall.

To our complete surprise, the two girls that we asked, accepted our advances and not only that, at the end of the dance we were invited to join their group. When they asked us where we were from and I told them England, they misunderstood and thought I meant New England, which of course was several thousand miles to the North. I had to explain that we were from the England over the sea and genuine Limeys. It transpired that we were the first English people that they had come across; consequently we were made most welcome.

We soon discovered that we had gate crashed the annual dance and prize giving of the local deep sea game fishing club, by the time it all ended, we had made many new friends. One in particular who went by the name of Jed invited us to visit his performing dolphins. He collected us the next morning in a battered old open top Cadillac and drove us for a mile or two out of town to a muddy looking creek.

He told us that he took the pair of dolphins which were named Pete and Flipper, around the country and did shows at venues that had suitable water areas. After parking the car we had to walk about fifty yards to the waters edge, where Jed threw a couple of stones into the creek. Immediately a pair of dolphin's heads popped up and came towards us. We fed them some fish that Jed had bought with him and they started to do all sorts of leaps and twists, all the time making squeaking sounds as though they were talking.

I noticed that they were in a crudely wired in compound, but were quite capable of leaping over the wire and swimming off. It was the first time I had seen dolphins at close quarters and I was impressed by their kindly nature and playful ways. However having seen hundreds of them in their natural environment, I felt it wasn't right to keep them under such conditions and told Jed what I thought. He understood what I meant, but seemed to think that they had become reliant on him for their food and would find it difficult to fend for themselves out in the ocean. It was obvious that they could swim off any time they wanted, but as they were getting a regular supply of food, they were quite happy to hang around.

The night before we sailed for Philadelphia, we invited our Texan friends down to the ship for a few drinks. They had never sampled Guinness before, so when the time came for them to leave, most of them were legless. I wouldn't have been surprised to hear that Jed's old Cadillac acquired a few more dents on the way home.

As I had been to Philadelphia before, I knew that it didn't have much to offer the likes of us, but the good news was that after Philadelphia we were going up to Newark, New Jersey where we would be able to visit the officers club in Manhattan. After this we were bound for the island of Bermuda, this was a bit of a detour as it meant a three day run out into the Atlantic, before going back up to Boston then up the Canadian coast.

With winter setting in there was some concern that the harbours up in the Gulf of St. Laurence would be icing up, with this in mind and to help speed things up we had the shore gang aboard while in Newark to give us a hand to change the piston rings. This was to make sure the main engine was up to scratch for the long haul back to New Zealand.

As we knew we still had many months of the voyage ahead of us, we decided to chip in and buy a record player. We bought it from the same shop that the Junior and I had a spot of trouble in when he purchased his tape recorder on our first trip. This time, knowing the ropes we produced our seaman's cards and managed to avoid paying the state tax without the police being called.

In the past when we held an engineers party in the smoke room we had to borrow a tape recorder from the Sparks, so we thought that having our own record player would make us independent and give us an edge with the girls when we arrived back in New Zealand.

We knew from experience that it would not suit our electric supply, but Ray was confident that he would be able to convert it to our needs. As soon we arrived back on board he disappeared with it down to the Lecky's workshop. After a few days he presented it in the smoke room in perfect working order. Records in America were much cheaper than back home, so we soon acquired a good selection that we knew would be suitable for ships parties.

Our shore gang turned out to be the same crowd that we had on board during my previous visit so we got on well together from the start. We also took the opportunity to send ashore four spare generator piston and con rod assemblies to have their top ends rebushed. This was something that we usually did ourselves, but as we didn't have the luxury of a hydraulic press the bushes had to be driven out using a 56lb hammer and a dolly. When using this method it was important to have a solid base to work on. Down in the engine room it wasn't possible to find anywhere really suitable, so when we carried out this operation we found the best place to do it was up at the after end of the upper deck on the redundant anti-aircraft gun mountings that were still in place from the war.

During one lunch time we heard on the radio that the President had been shot. They say that everyone remembers where they were when J.F. Kennedy was assassinated and this was certainly true in my case. I well remember the look on the face of the Forman of our intrepid shore gang when I told him about it after lunch. He didn't believe me at first, but when he realised it was true, he told the rest of his lads and they all downed tools and went home.

That night we took the bus into Manhattan and found the streets including Time Square, practically deserted. By this time it had been confirmed that the President was dead so most of the bars and taverns of ill repute had closed. We made our way to the Great Northern Hotel and the Officers Club but found a very subdued atmosphere pervading. After a short while we decided to head back to the ship and call it a night.

The next couple of days were very depressing, so we were glad when the time came to leave for Bermuda. Unfortunately we found the island very expensive and being late November it was very quiet with a definite shortage of the fairer sex. There were some nice beaches near our berth in Hamilton, but unfortunately the sea temperature was not conducive for taking a dip, even though it was crystal clear and looked very inviting. If we had been a couple of months earlier, we would have been able to have drooled over and perhaps tried our luck with the bikini clad daughters of the rich American holiday makers that frequented the place during the Summer.

There were one or two reminders of home about the place; like the very English post boxes and the English looking Bobbies on the beat. The two that stay in my memory, were the sight of British made cars and the unaffordable price of a pint of Whitbread's best bitter, which was around seven times as much as it cost back home.

We stayed in Bermuda for two or three days before heading back up to America and the port of Boston where we stayed just long enough to discharge the last of our frozen meat. As the weather was getting colder by the minute, our shore excursions were rather limited due to our not having any warm clothing.

One of the things that I remember about Boston was the local bus service. You paid ten cents for a ride, whether it was for one stop or all the way, so this was a cheap way to see something of the place. When we last visited Boston it had been in the summer so I had been able to take advantage of the nearby beach and eye up the local talent. The beaches were now deserted and all the girls had gone to ground so we were glad to be on our way.

We left Boston on a very cold and frosty morning for Halifax, so the warmth of the engine room was very welcoming. With the sea temperature down to almost zero, the oil pressure on all our machinery was nice and high which kept our troublesome generators happy and free from any serious problems that were the norm in tropical waters.

With Christmas approaching, the main topic of conversation was whether we would be spending it at sea or in port. We continued to head north to the Gulf of St. Laurence; where we were due to take on a consignment of newsprint from Port Chandler. Originally we were going up to Montreal but as the St. Laurence River was iced up we were diverted to the little port of Chandler. This turned out to be at a remote jetty that tested the manoeuvring skills of the lads up on the bridge.

Sometimes when the bridge rang down manoeuvres in continuous quick succession we would phone them up and tell them that we were running out of compressed air and they only had three starts left. This was not necessarily true but it did make them realize that just moving the telegraph handle didn't always mean the ship would respond in the way they expected.

On this occasion there was a series of panic like movements and the Chief Lecky was having trouble writing them down quick enough in the movement book. We were becoming concerned that our compressors wouldn't be able to keep up with demand and the pressure gauge was dropping like a lead balloon, so we had to call the bridge and make them aware of the situation. As it turned out the approaches to the jetty were badly charted and there were blizzard like conditions prevailing up top, making docking a very difficult undertaking and I was told later that we actually went aground at one point.

The following day our Third Mate, Peter King organised a boat crew from the Cadets. Unfortunately the engine in our crash boat wouldn't turn over; this was because the flywheel had iced up to the bell housing. The engine being air cooled had the fan built into the flywheel so there was minimal clearance between it and the casing. Peter somehow borrowed the harbour masters launch and sounded the whole area and took bearings before returning to the ship and preparing a very accurate chart. Cartography was Peter's hobby and I seem to remember that a copy was then presented to the Harbour Master; so I wouldn't be surprised to hear that it is still in use today.

It was getting close to Christmas when we heard that we were going across the Gulf to Corner Brook in Newfoundland. This was to load the last of our cargo destined for New Zealand. As it was only a days run, it now looked pretty certain that we would be spending our Christmas there.

After a rough crossing we negotiated Newfoundland's Bay of Islands and continued up a wide inlet with a formidable mountainous coast line. I was on watch when we docked and it was at times like this I was pleased to be in the comforting warmth of the engine room. I was glad that I wasn't up on the bridge with all the responsibility of trying to safely navigate this fearsome coast, with frequent periods of zero visibility caused by freezing fog. The ice built up on the rigging and superstructure up forward and gave the ship a ghost like appearance.

When I went topside after 'Finished with Engines' was rung down I found the quayside and surrounding buildings covered in a thick layer of snow, what a contrast to not many days before when we were in Bermuda. I was surprised to see berthed astern of us the *Nicholas Bowater*. She was the largest (7136tons gross) of Bowater's fleet of seven fine ships. Their usual run was back and forth across the Atlantic with paper pulp. It was not something that appealed to me, but no doubt it suited those that wanted to be home on a regular basis.

I had heard that Sid, one of my fellow apprentices at the London Graving Dock Company had gone to sea with Bowater's. He had got himself married towards the end of his apprenticeship, so it made sense for him to join their fleet, especially as we used to do quite a lot of work on their ships at their paper mill at Northfleet on the lower reaches of the Thames.

During lunch time I paid a visit to the *Nicholas Bowater* in the hope that he might be on board. I found a group of engineers having a lunch time hydraulic session in the

Seconds cabin, so I introduced myself and enquired if Sid was aboard. Unfortunately no one had heard of him, so it seemed unlikely that he was still with the Company or in fact had ever been with them at all.

I was handed a beer and we started swinging the lamp, when I told them that I had worked on their ship when I had been with the London Graving Dock they welcomed me almost as a Company man. Apparently the Graving Dock had a good reputation with Bowater's and at the time still carried out work on their ships. They jokingly said that I had probably seen more of the inside of their engine than they had. There might have been an element of truth about that, as the Amalgamated Engineers Union (AEU) didn't allow ships engineers to do their own repairs while in the UK.

The *Nicolas Bowater* was powered by a steam turbine and at the time I worked on her in the late fifties she was almost a new ship. Any work on her turbine would have been carried out by the 'Makers Men'. The work that I had actually been involved with was removing for overhaul some steam valves

Berthed in Corner Brook, Newfoundland with the *Nicholas Bowater* astern. (Photo John Layte).

on top of the boilers, these were sent back to our machine shop at Poplar. We then refitted them after they had been serviced.

The beers continued to flow so it was about 3.00 pm when I staggered back through the snow to the *Rakaia* with the good news that we had all been invited to a Christmas Eve party on the Nicholas Bowater. It turned out to be a terrific do with plenty of girls from some sort of theatrical society. The Bowater boys were regular visitors to Corner Brook so they had all the right connections. It was felt that we couldn't let our side down, so we invited them all over for a party on the *Rakaia* for the evening of Christmas day.

Christmas morning I woke up to find that I had a Christmas present; when I opened it I found it was a hand knitted scarf and a pair of gloves. The card that came with it said happy

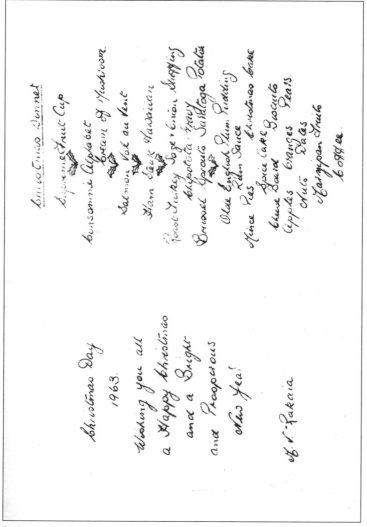

Christmas day dinner menu.

Xmas from the girls at the Officers Club NY. It was the only time that I had a fleeting feeling of home sickness during my time at sea. It transpired that we all received similar practical presents from our good friends at the club.

On the way up the coast the main engine had developed an occasional strange noise from the area around the scavenge blower casing. The Chief dug out the drawings and it was concluded that it was probably caused by slackness in the Bibby type drive coupling between the huge chain drive sprocket and the blower shaft. As it was Christmas he said it could wait until Boxing Day before we dealt with it.

Our Christmas dinner was a superb affair, Gerry and the rest of his galley staff excelled themselves, I am quite certain that it couldn't have been bettered in any top class restaurant. After dinner we set up the smoke room in readiness for the evening do. However Gerry and his team hadn't quite finished their endeavours, when I looked in the dining saloon, the tables had been arranged into one huge one that was covered with artistically prepared dishes that resembled a medieval banquet.

Usually when there was party in the smoke room, it was a Mates only, or Engineers only affair. There was a minor rivalry that seemed to exist in the N.Z.S.C. between us. I think this was probably due to the way the accommodation was designed. The Mates occupied the upper deck while we had the port side alleyway on the boarding deck; this led to a natural segregation.

Later when I sailed with Buries Markes, my cabin was just around the corner to the Second Mates, so the question of demarcation didn't arise. The Second Mate at that time was known as the 'Red Rooster' on account of his beacon like nose. He was a great character and kept a large apple tree in a tub in the corner of his cabin and was convinced that it would produce fruit. This was because he said that he had seen a bee on it while we were in the dry dock at Falmouth. He also kept a pair of giant African land snails that lived mainly on Guinness and the leaves of the apple tree.

They were what would be termed 'free range', so you had to be careful when entering his cabin.

We decided that as it was Christmas and we were expecting a surplus of girls, we would invite the Mates along to our party to help us out. In the event the night was a roaring success, the Bowater boys said it was the best party they had ever been to. Before the girls left in the early hours of the morning they invited us along to a do that evening at the theatrical societies H.Q., which of course was Boxing Day.

However there was the slight technical hitch in that we had to repair the blower drive before we could go. The job had to be completed that day as we were due to sail for Panama on the afternoon of the 27th. Despite enormous hangovers, we set to after a couple of hours sleep and worked right through until the job was boxed up in the early evening.

The Mates had already left, so not to be outdone, we skipped our dinner and hot footed it down the gangway to a pair of waiting taxis and sped off to the sound of rattling snow chains and arrived just as the party was in full swing. Unfortunately we had to leave the Fiver behind as it was his turn to be duty engineer.

It proved to be a great night, but you can only burn the candle at both ends for a limited time, so we were pleased to get underway again, with the prospect of sunshine and blue seas not many days ahead.

24
Fire Down Below

The first couple of days were spent in a trance like state; luckily the repairs to the blower drive solved the occasional knocking sound that we had experienced for the last few miles up the Newfoundland coast and the rest of the machinery was running well. It was a nine day run down to Panama and three and a half days out we crossed the two tracks that we made on our trip out to Bermuda, so forming our own Bermuda triangle.

We were now back in the blue water and the return of our deck activities was welcomed by all. On the sixth day we went through the Windward Passage between Cuba and Haiti and entered the Caribbean. Two and a half days later was a high point, as we collected our mail at Cristobal before starting our transit of the Panama Canal. Everything was running well as we left Balboa astern and entered the Pacific and the cool waters of the Humbolt Current.

We were now working hard in an effort to regain our tans

Regaining our tans after being confined to the accommodation during our semi Arctic wanderings.

that had been somewhat diminished over the last few weeks of our semi Arctic wanderings. We still had twenty one days to perfect them, so that they could be used to advantage with the young and sometimes not so young female population of New Zealand. On this Pacific crossing we were going direct to Auckland, so it was unlikely that we would be seeing any land before we reached the coast of New Zealand.

Two days out from the canal the ship cleared the Humbolt Current and immediately the sea temperature shot up necessitating our coolers to be fully opened up. We started our standby lub oil pump to the main engine and before long both pumps were working flat out to maintain a reasonable pressure.

During my 12 to 4 graveyard watch and about a week into our crossing, Kiwi our Sixer informed me that there were fumes coming out of the oil filler on number one generator. At the time I was finishing off working on number three generator. Number four geny was also out of action waiting for some T.L.C., so as I only had a couple of valve clearances to set before barring it round to its starting position, I told him to keep an eye on it and let me know if things got any worse. I realised that something was seriously wrong, but I had no choice, but to hope I could get number three up and running in time.

Kiwi returned almost immediately and shouted that the engine room was on fire. I realised that I had no chance of completing the job in hand and rushed round to assess the situation. In the short time that Kiwi had taken to come round and give me the news, things had drastically deteriorated.

As I rounded the front of the of the main engine my feet went from under me and I slid along the plates on a surface of hot oil and water into a shower of sparks being thrown off the brush gear of number two generator.

It's surprising how quick you can take in your surroundings in an emergency. The scene still remains in my memory, just as though I had taken a photograph of it at the time. There was a fountain like stream of oil and water shooting out of the oil filler of number one generator and it was falling onto the covering plate of the commutator and brush gear assembly of number two generator. It was then being sucked on to the brushes where it caught fire; causing a Catherine wheel effect of sparks that in turn set fire to the oil that covered the engine room plates and the crankcase door that was in the line of fire.

Engine room fires are serious business, so there was no way that Kiwi and I could cope with the situation on our own. As I glided through the flames on my backside, the old saying, 'out of the frying pan into the fire' came to mind. I came to a halt right next to the controls and regained my feet and pressed the emergency alarm. I then brought the main engine to a stop and shouted to Kiwi to shut down number one generator while I did the same for number two.

A stunning silence suddenly pervaded the engine room, but there was no time to contemplate it, the paint on the side of the main engine was now well alight and so was an area of the bilges under the failed generator. The flames gave us enough light to grab a couple of extinguishers and give the fires a good dose of foam, this immediately killed the flames but plunged us into darkness. The whole disastrous episode had had only taken just over a minute from start to finish. With all the machinery running flat out and each part reliant on the rest, when something went wrong it could easily become a failure to proceed situation.

I could see the beams of torches sweeping about above us as the rest of the lads made their way down. I positioned myself at the bottom of the steps so that I could warn them about the oil that covered the plates; the first to arrive was the Second.

"Looks like the shits hit the fan this time Davy boy," he remarked as he removed the wet rags from around his head.

"Its not as bad as it looks, the main problem is with number one, if we go round to number three, we should be able to get it going in a couple of minutes and that should give us enough power to get the pumps going and get the main engine under way," I answered.

We soon had the main engine running on reduced revs and once underway we were able to access the situation. The

Lecky's set about overhauling the brush gear and cleaning up the commutator on number two, so that by the end of my watch it was back on load and the main engine was back up to normal revs.

Number one was a different matter; we still had to find out why it had lost most of its oil and filled its crankcase up with cooling water. Once we removed the crankcase doors we found that one of the cylinder liners had split open allowing the cooling water to poor in and force the oil out and over the armature of number two which centrifuged over everything.

The paint on the crankcase door of the main engine that was in the firing line of the burning oil from the Catherine wheel effect was completely burnt off, so we were lucky not to have had a major engine room fire.

We went on to six hour watches until we had three of our four generators all in running order. We fitted a new liner and reconditioned piston assembly to Number one and by the end of the following day it was ready for a test run. We had been lucky with the bottom end and main bearings; the only bearing that we had to replace was the bottom end to the offending unit. We arrived in Auckland with no further incidents and the lads hot footed it into town to visit the usual haunts.

As it was docking day I had to stay on board as duty engineer, so I missed out on all the fun. Perhaps being promoted to Third engineer wasn't such a good idea after all. Our routine maintenance commenced the next morning with changing piston rings and checking the bearings on the main engine. The lecky's were kept busy overhauling the brush gear on all the main engine pumps and checking over the steering gear and the Freezers were fully occupied bringing down the temperatures in the holds.

As we were all fully employed we didn't bother to get scrubbed up and changed for lunch in the saloon. Depending on how dirty we were, we either had our lunch in the mess room or in one of our cabins. We had been in Auckland for about a week and six of us were having a mainly hydraulic lunch in my cabin when smoke started drifting through the doorway. We rushed out to find the alleyway full of oily smelling smoke, luckily someone had left the engine room door open, otherwise we wouldn't have been any the wiser.

We piled down the engine room to find an exact scenario of what had happened a few weeks previously. Number one generator had spewed out its oil onto number two's commutator and the whole lot was on fire, including the freshly painted crankcase door which the Donkyman had done to a coach painter's standard. I started one of our spare generators and the Fourth started the other while the rest of the lads stopped the offending one and put out the fire. We soon had it all sorted and under control so we returned to my cabin to resume our lunch.

We had just opened some fresh cans when we heard the fire brigade arrive. Then the Fourth Mate put his head into the cabin and shouted 'Fire' and immediately raced off down the alleyway, screaming,

"The engine rooms on fire and all the engineers are all on the piss."

The next thing, several firemen ran past the doorway, the last of them was unrolling a large reel of fire hose.

Kiwi looked out and said that they had all gone down the engine room. Admittedly there was still a lot of smoke

Third Engineers cabin, fuelling up prior to a shore side evening foray.
Left to right: 2nd Engineer, 4th Engineer, Engineering Cadet, Junior Engineer, 5th Engineer, 3rd Engineer (Author).

drifting out of the engine room, but we knew that there was no fire as we had already dealt with it. The Fourth mate came back about ten minutes later and told us that everything was now under control. Kiwi handed him a beer and informed him that we had put out the fire before he even knew it had started and as he had called the fire brigade he would probably have to stand for the bill. We all had a good laugh, finished our drinks and went back to work.

After a week in Auckland we continued our loading program around the various ports on the New Zealand coast. It was now midsummer, so the beaches were full of beautiful New Zealand girls who were only too pleased to accept our advances and invite us to their parties in the evenings.

It was not uncommon to return to a port that we had visited two or three weeks previously, however there were certain disadvantages with returning too soon, as I had discovered on my last trip. The image of the two gorillas that were intent on preserving their sister's reputation was still fresh in my memory, so it was advisable to make sure you hadn't blotted your copybook on the previous visit before venturing ashore.

The only port that we visited twice on this occasion was Napier and as we always met up with the same crowd of girls, it was very unlikely that we would have any unwelcome male guests storming the ship. Sadly our stay came to an end, but fortunately the loss of our romantic notions was eased by the knowledge that our next and last port of call was to be Nelson at the top of the South Island where the *Rakaia* was adopted.

We docked in Nelson to the sound of the local pipe band and as all our major engine room repairs were finished, it looked like we would be having an easy time for our last week in New Zealand. The ship held a Cadets and Officers party which was a roaring success and got us off to a good start. Two days before we were due to sail for home the nurses at the local lunatic asylum invited us to a party at the hospital. They seemed to have a particular affinity with us engineers, so their functions were always something to look forward to!

We were down below on the day before the party when Number one generator suddenly started to emit smoke and steam out of its oil filler aperture. This time we were able to stop it before it disgorged its oil over number two and set the surroundings on fire. It was now obvious that there was something causing the reoccurring problem that we hadn't spotted.

It so happened that the Company's New Zealand Superintendent was on board at the time and he was quite adamant that the crankshaft must be cracked or the main bearings were out of line. However the Chief didn't agree as we had taken crankshaft deflections after the two previous occasions and they were all found to be satisfactory. He came to the conclusion that the cause of the trouble must be in the top end bearings of the piston assemblies that had been sent ashore while we were in Newark.

We still had one left, so we took it up to the old gun platform which was above the Chief Freezers cabin and knocked out the gudgeon pin. At the time Eddie the Chief Freezer was ashore with his drill bit and thermometer, checking the internal temperature of random carcases that were about to be loaded. But the next day I heard him asking Chippy Newlyn if he would arrange to have the deckhead in his cabin repainted as a great sheet of it had somehow fallen off on to his bunk.

After checking the clearance between the top end bush and the gudgeon pin we found that it was only three thousands of an inch instead of the eight that it should have been. What was surprising was that the so called reconditioned ones had run for as long as they had. The fact that each of them had run for over a hundred hours before overheating and finally resulting in Eddies bunk being covered in old paint could only be put down to the 'Law of Unintended Consequences'.

As we were due to sail the next day, we had to turn to with a vengeance and hope we could solve the problem before the nurse's party that evening. The Super was only concerned with getting the ship to sea on time and pass the responsibility of the cargo on to the Chief. With the need to have all four generators running for most of our run home, the chief on the other hand made it plain that he was not prepared to sail until the faulty generator was up to it.

233

To save time it was decided to fit the best of our collection of used piston assemblies instead of drawing out the top end bush and boring it out on the old lathe which in itself was a museum piece. This was something that could be done once we got to sea. By the time afternoon 'Smoko' came round we had the engine boxed up and ready to run.

By dinner time that evening it had been running on normal load for two hours and seemed fine. Just before going into dinner the chief called me aside and told me to nip down and increase the load on number one to 900 amps, then come straight back up. This was effectively a 50% overload. As soon as I eased the load off the other running generator and put it on number one the engine started to protest, I didn't hang around, but hot footed it up to have my dinner.

I had just entered the accommodation when I encountered the Super on his way to the engine room.

"Everything O.K. down there then Third?" he asked.

"Seems to be sir, everything was running O.K. when I came up."

But I was also thinking that shit was going to hit the fan, if he went down to check for himself.

I carried on into the dining saloon and got tucked into my Mulligatawny soup, we always had this the day after one of Gerry's red hot curries. I was on the point of finishing it when the Super stormed in and confronted the Chief, saying that number one was on overload and was about to seize up because I had tried to sabotage it.

Everyone held their breath, thinking that the Chief would explode, but instead he just placed his knife and fork on his plate, looked up and calmly said,

"I'll trouble you not to discuss engine room business in the dining saloon; I'll look into it in a wee while. But for your information and for the record, I instructed my Third Engineer to increase the load to 900amps. If the engine can't take the load that it was designed for, then I am not prepared to take the ship to sea. I need to be sure before I inform the Captain here that we are seaworthy."

A deafening silence descended in the saloon as we all waited to see what would happen next. The Chief calmly picked up his knife and fork and continued with his meal, while the Super realising that he had overstepped the mark made a silent retreat. I expected the old man to make a remark but he wisely decided that on this occasion, discretion was better than one upmanship.

After we had all finished our dinner and retired to the smoke room, the chief called me aside and said the Second was going below to check things out and would I go with him. If number one was holding its own, he told me to bring it back to 600amps and leave it running. As we descended the steps, I could just make out it running through a thick fog of hot oil mist that was rising off every part of the engine. We went round it and checked all the temperatures and had a good listen with our stethoscopic screwdrivers for any untoward noises emanating from its internal organs. The Second was satisfied with it so I went up to the board and transferred 300amps on to the other generator. The good news was that we were now ready for sea and even more ready for the nurse's party in their quarters out at the hospital.

We arrived back on board in time for breakfast and much the worse for wear. During 'Smoko' it was generally agreed that although we had all had a great time ashore, what we needed was a rest, so in a way we were looking forward to getting back to shipboard routine for the five week voyage home. There were a few 'just jobs' that had to be seen to before sailing time but they were all completed by the time I took the first watch at midday.

Just before 'Stand by' was rung down the bridge phoned and requested that we carried out the usual stowaway search, which we did by waiting a few minutes before phoning them back informing them that no stowaways had been found. Almost immediately 'Stand by was rung down and as it was my watch I took the controls. Anticipating that the first movement would be 'Slow Astern', I had the reversing lever already in position. Once the ship was sprung off the quay, 'Half Ahead' followed and we were away.

Most of us were down below at the time, but I heard that there was the usual crowd of well wishers on the quayside to see us off as a visit of the *Rakaia* was quite an event on the Nelson calendar in those days. How different from back home where the arrival of ships with vital food supplies from the far side of the world was taken for granted by the general public.

25
Back to School

On our homeward run we were due to stop at Pitcairn Island for two or three hours. This would break up the long haul across the pacific and if necessary give us time to do the odd minor repair, such as change a suspect fuel valve on the main engine or service one of the generators while we had some spare amps. We would also be able to stock up on fresh fruit, my favourites were the islands grapefruit, they were so sweet that you could eat them like an orange and I have never tasted better anywhere. We would also take on the usual complement of biting insects that would keep us scratching for a few days and perhaps give a change of diet to our colony of flesh eating cockroaches.

With all the trading completed we waved the islanders off and as they pulled away singing 'In the Sweet Bye and Bye' we went below and got underway. We had good weather and no mechanical problems all the way to Curacao where once again we took on fuel.

Surprisingly, number one generator was running well, as were the other three, so apart from the odd scavenge fire we had the easiest voyage home that I had experienced. By the time we reached the Western Approaches we had been together for nine months. This was usually the time when the 'Channel Fever' showed its first symptoms among the crew, but something seemed to be different. It had been a long voyage and we had become a close nit community, especially over the last few months. There seemed to be an unspoken feeling of regret that we would soon be leaving the ship and going our separate ways.

Our record player had done sterling service since we bought it back in Newark and it had definitely enhanced our shipboard parties during the weeks on the New Zealand coast. We didn't particularly want to leave it behind on the ship so we raffled it off among the crew and split the proceeds between those of us who had chipped in to buy it. The unfortunate winner was our seventh engineer; I say unfortunate because he would have to add this to his luggage and the copious amount of junk that we all seemed to acquire during a voyage.

We entered the locks to London's Royal group of docks on the early morning tide and made our way with the assistance of a tug up to our berth at the far end of the Royal Albert Dock. It was springtime but the London grime and drizzle made it all look very depressing after our blue water voyaging, nevertheless it was a welcome sight.

Our trip home had been remarkably trouble free; our earlier troublesome number one generator was still running like a sewing machine when we shut it down after 'Finished with Engines' was rung down. As usual at the end of a voyage the Chief Superintendent Engineer, Mr Strachan came aboard and interviewed each of us as to out intentions.

I hadn't really given it much thought, but suddenly here we were once again after a voyage of nine months and five days and everything feeling unreal. When it came to my turn for a chat with the Super I was undecided about my future plans. The Chief was retiring, so the *Rakaia* wouldn't be the same without him, therefore I thought I would request a transfer to one of the Company's other ships.

The Super said that both the Rangitoto and the Piako were due back in the U.K. and I could have the Thirds job on either of them. I didn't fancy the Rangitoto as being a passenger ship certain rules of etiquette were more pronounced than on cargo ships within the Company. The Piako on the other hand sounded very good, she was only two years old and powered by an eight cylinder Sulzer main engine.

The Super didn't give me time to consider my options.

"What I recommend," he said, "is that you go back to college and sit for your Seconds ticket, you have had plenty of experience now and you have certainly got your sea time in."

I hadn't given this option any thought up until then, but after consideration I decided to take up his recommendation.

It was agreed that I stand by the *Rakaia* until she was due to sail and then take what was known as ticket leave. This was two months paid leave to help pay for my college studies, after that I would be entitled to go on the dole for the next thirteen weeks to see me through most of the course. I hoped that together with my pay off money of three hundred and twenty seven pounds twelve shillings and sixpence, it would be enough to see me through before financial cramp set in.

I transpired that most of the lads were being allocated to new berths, the only ones returning were Brian (The Bear) he was going to be sailing as the new Chief Electrician and Bob our junior engineer, who I understood was being promoted to Second Freezer. We paid off that afternoon and we said our goodbyes and I made the short journey home, then went round to my Aunts garage and recommissioned my old Rolls Royce.

The next morning I returned to the ship, now that nearly everyone had left it seemed completely different. It made me realise what made a happy ship, it was all down to the crew and its leadership. A standby Chief Engineer was on board and a junior was due to arrive. I collected most of my gear including my duty frees plus a little extra, then went and found Eddie the Chief Freezer who was handing over the responsibility of the frozen cargo to his replacement. He had been looking after a case of New Zealand apples and a frozen lamb carcass for me and as soon as the standby engineer arrived I loaded it all into the Rolls and set off for home.

While working in the docks during my apprenticeship I learnt the odd skill quite unconnected with the engineering side of things, so consequently I was waved through the dock gates by the Port of London Authority (PLA) policeman on duty with no trouble at all. Once home I took the lamb carcass round to our local butchers as usual and after he had taken his pound of flesh, my family had succulent New Zealand lamb on the menu for the next few weeks.

After standing by the ship overseeing the shore gang from R.H. Green & Silly Weir dismembering our main engine and various other bits and pieces, the time came when the sailing engineers joined the ship. I had to vacate my cabin for the new Third, but as I had now joined the nine to five brigade and was sleeping at home for most of the time it was of no consequence.

Suddenly my time on the *Rakaia* came to an end and I left the ship as the Blue Peter was being run up the mast a day before she was due to sail for Australia. To anyone with an inexperienced eye taking a look down the engine room, they would have been convinced that the ship would never be ready in time. As usual the last twenty four hours before sailing the conditions down below gave the impression of a ship breakers yard rather than a ship about to leave for the other side of the world.

In hindsight the few years that I sailed with the New Zealand Shipping Company were some of the happiest of my life and some of the friendships made, have stood the test of time, but unfortunately some of my old shipmates have now crossed the bar. What none of us realised at the time was that we were the last generation of hands on engineers.

Within a few years the whole world order of trade changed, the U.K. deserted its old Colonial food supplies in favour of the common market and the way of life of most British deep sea seafarers changed forever. Containerisation was rearing its ugly head and beautiful looking ships like the *Rakaia* were beginning to disappear from the oceans of the world. The dockside haunts favoured by seamen were soon to be replaced by up market wine bars, frequented by people that belonged to the mutual admiration society, who had no conception of what it was like to do a proper days work.

Durham Association Lunch at Liverpool's Maritime Museum 2008

Paul Wood Ex. N.Z.S.C. Cadet, Dave Carpenter Ex. N.Z.S.C. Engineer, Brian Anderson Ex. N.Z.S.C. Electrician.
All shipmates on the *Rakaia* for voyage 36 (Double Header).

The *Rakaia*, like the rest of the N.Z.S.C. fleet were absorbed into the Federal Steam Ship Company and adopted their livery. Cargoes were getting hard to find on an economical basis and within a few years she ceased to be a cadet training ship and for a few months she was laid up in the River Fal, which was the nautical equivalent of an old people's home. She was sold to the ship breaking firm of Lee Sing in 1971 and demolition began in Hong Kong the same year. At least she didn't have to suffer the indignity like the rest of the fleet, of becoming part of the P&O cargo division and having her funnel painted in the awful livery of that company.

I returned to Poplar Technical College for my studies, where I was greeted by one of my old lecturers with, "Back again Mr Carpenter, what a pity they don't put old heads on young shoulders."

However two months into the course with the lack of filthy lucre rearing its ugly head, I obtained a great deal of satisfaction by seeing the look on the normally expressionless faces of the staff at the Social Security office in Stepney, when I pulled up in the Rolls to collect my dole money of two pound seventeen shillings and six pence. As it was a Wednesday I collected my copy of Lloyds List from the newsagents in Poplar High Street, for some reason I opened it at the situations vacant page.

Third Engineer wanted for British registered motor ship for voyage to the Far East.

Looks interesting, I thought, I might give them a call!

Still Game (2010).

Appendix I

Owing to the complicated design of *Rakaia's* main engine a brief description is given below.

Each unit (Cylinder) has three pistons, ie. One exhaust piston at the top, the main piston in the middle and an exhaust piston at the bottom.

The exhaust pistons are attached to cast steel yokes.

The yokes are connected together by four tie rods, one on each corner.

The bottom yoke has two rods that are attached to their own eccentric strap that fits to an eccentric that is fixed on either side of the main crankshaft webs. Consequently both the exhaust pistons rise and fall in tandem.

The main piston works in opposition to the top and bottom exhaust pistons, thus giving a combustion chamber above and below the main piston. The main piston rod passes through the centre of the bottom exhaust piston; combustion pressure is sealed by a specially designed stuffing box.

The main piston rod is attached via a crosshead to the connecting rod which is operated from the crankshaft journal.

The scavenge air is supplied by rotor type blowers driven by a duplex chain of 2.5 inch pitch.

An exposed camshaft mounted on the Starboard side operates the fuel pumps and is driven by a single 4 inch pitch chain from the crankshaft. It is also used for taking indicator cards.

It's interesting to note that with an 8 cylinder engine, such as the *Rakaia's* there are 16 combustions for every revolution.

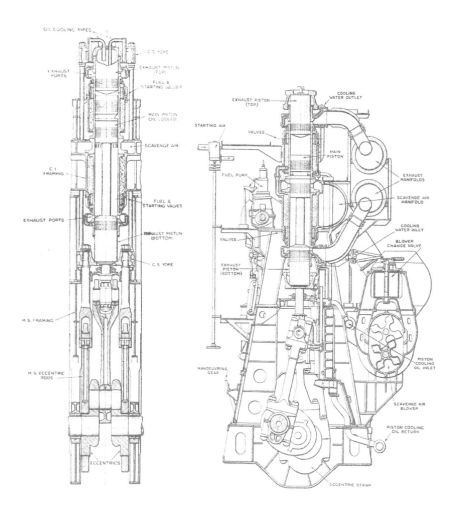

Main Engine
Section of one Unit

Main Engine
Looking Aft

246

Appendix II

A brief description of the M.V. RAKAIA

Built by Harland & Wolff and launched at their Belfast yard on the 30th December 1944.
Delivered to the Ministry of War Transport as the EMPIRE ABERCORN on the 30th June 1945

Tonnage	8213 gross
Length	474.2 feet (overall)
Beam	63.3 feet
Draft	29.2 feet (summer)

Sold to N.Z.S.C. in 1946 and registered in London as RAKAIA
Converted to Cadet training ship in 1958
Ceased in this role in 1968
Broken up in Hong Kong 1971

Main Engine
Burmiester & Wain (Direct Reversing)
8 Cylinder double acting opposed piston two stroke diesel.
7500bhp @ 120rpm

Cylinder Bore	550mm
Main Piston Stroke	1200mm
Exhaust Piston Stroke	400mm

Built by Harland & Wolff to a design by C.C. Pounder
Speed 14.5 Knots @ 102 rpm

Total Fuel consumption 29 tons gas oil / 24hrs
Total bunkers 1411 tons

Generators
Four 6 cylinder Harlandics
250KW – 220 Volts
335 bhp @ 420 rpm

Boiler
Harland & Wolff Clarkson cylindrical
7 foot 11 inches diameter
20 foot 9 inches high
Working pressure 100 Ibs/sq in
Fuel: gas oil or waste heat

Refrigeration Machinery J & E Hall (Dartford, Kent)

Three units   Harland & Wolff motors 550 amps 220 volts
Six compressors
Normal ships compliment 96 souls

Further information on the M.V. *Rakaia* and other cadet training ships of Great Britain's Merchant Navy can be found by visiting:

www.rakaia.co.uk
www.shipsnostalgia.com